AFOQT Subject Test

Mathematics

Student Practice Workbook

+ Two Realistic AFOQT Math Tests

Math Notion

www.MathNotion.com

AFOQT Subject Test – Mathematics

AFOQT Subject Test – Mathematics

AFOQT Subject Test Mathematics

Published in the United State of America By

The Math Notion

Web: WWW.MathNotion.com

Email: info@Mathnotion.com

Copyright © 2021 by the Math Notion. All rights reserved. No part of this publication may be reproduced, stored in a retrieval system, or transmitted in any form or by any means, electronic, mechanical, photocopying, recording, scanning, or otherwise, except as permitted under Section 107 or 108 of the 1976 United States Copyright Ac, without permission of the author.

All inquiries should be addressed to the Math Notion.

ISBN: 978-1-63620-039-2

AFOQT Subject Test – Mathematics

The Math Notion

Michael Smith has been a math instructor for over a decade now. He launched the Math Notion. Since 2006, we have devoted our time to both teaching and developing exceptional math learning materials. As a test prep company, we have worked with thousands of students. We have used the feedback of our students to develop a unique study program that can be used by students to drastically improve their math scores fast and effectively. We have more than a thousand Math learning books including:

- **GED Math Prep**
- **HiSET Math Prep**
- **TABE Math Prep**
- **TASC Math Prep**
- **ASVAB Math Prep**
- **many Math Education Workbooks, Study Guides, Practice and Exercise Books**

As an experienced Math test preparation company, we have helped many students raise their standardized test scores—and attend the colleges of their dreams: We tutor online and in person, we teach students in large groups, and we provide training materials and textbooks through our website and through Amazon.

You can contact us via email at:

info@Mathnotion.com

AFOQT Subject Test – Mathematics

Get the Targeted Practice You Need to Ace the AFOQT Math Test!

AFOQT Subject Test - Mathematics includes easy-to-follow instructions, helpful examples, and plenty of math practice problems to assist students to master each concept, brush up their problem-solving skills, and create confidence.

The AFOQT math practice book provides numerous opportunities to evaluate basic skills along with abundant remediation and intervention activities. It is a skill that permits you to quickly master intricate information and produce better leads in less time.

Students can boost their test-taking skills by taking the book's two practice AFOQT Math exams. All test questions answered and explained in detail.

Important Features of the AFOQT Math Book:

- A **complete review** of AFOQT math test topics,
- Over 2,500 practice problems covering all topics tested,
- The most important concepts you need to know,
- Clear and concise, easy-to-follow sections,
- Well designed for enhanced learning and interest,
- Hands-on experience with all question types
- **2 full-length practice tests** with detailed answer explanations
- Cost-Effective Pricing

Powerful math exercises to help you avoid traps and pacing yourself to beat the AFOQT test. Students will gain valuable experience and raise their confidence by taking math practice tests, learning about test structure, and gaining a deeper understanding of what is tested on the AFOQT Math. If ever there was a book to respond to the pressure to increase students' test scores, this is it.

WWW.MathNotion.COM

… So Much More Online!

- ✓ FREE Math Lessons
- ✓ More Math Learning Books!
- ✓ Mathematics Worksheets
- ✓ Online Math Tutors

For a PDF Version of This Book

Please Visit WWW.MathNotion.com

AFOQT Subject Test – Mathematics

Contents

Chapter 1 : Integers and Number Theory 11
    Rounding 12
    Whole Number Addition and Subtraction 13
    Whole Number Multiplication and Division 14
    Rounding and Estimates 15
    Adding and Subtracting Integers 16
    Multiplying and Dividing Integers 17
    Order of Operations 18
    Ordering Integers and Numbers 19
    Integers and Absolute Value 20
    Factoring Numbers 21
    Greatest Common Factor 22
    Least Common Multiple 23
    Answers of Worksheets 24

Chapter 2 : Fractions and Decimals 27
    Simplifying Fractions 28
    Adding and Subtracting Fractions 29
    Multiplying and Dividing Fractions 30
    Adding and Subtracting Mixed Numbers 31
    Multiplying and Dividing Mixed Numbers 32
    Adding and Subtracting Decimals 33
    Multiplying and Dividing Decimals 34
    Comparing Decimals 35
    Rounding Decimals 36
    Answers of Worksheets 37

Chapter 3 : Proportions, Ratios, and Percent 40
    Simplifying Ratios 41
    Proportional Ratios 42
    Similarity and Ratios 43
    Ratio and Rates Word Problems 44

WWW.MathNotion.Com

AFOQT Subject Test – Mathematics

Percentage Calculations...45
Percent Problems...46
Discount, Tax and Tip...47
Percent of Change..48
Simple Interest..49
Answers of Worksheets...50

Chapter 4 : Exponents and Radicals Expressions......................................53

Multiplication Property of Exponents..54
Zero and Negative Exponents...55
Division Property of Exponents..56
Powers of Products and Quotients..57
Negative Exponents and Negative Bases...58
Scientific Notation...59
Square Roots..60
Simplifying Radical Expressions...61
Answers of Worksheets..62

Chapter 5 : Algebraic Expressions..65

Simplifying Variable Expressions..66
Simplifying Polynomial Expressions...67
Translate Phrases into an Algebraic Statement...68
The Distributive Property..69
Evaluating One Variable Expressions...70
Evaluating Two Variables Expressions...71
Combining like Terms...72
Answers of Worksheets..73

Chapter 6 : Equations and Inequalities...75

One–Step Equations...76
Multi–Step Equations...77
Graphing Single–Variable Inequalities...78
One–Step Inequalities..79
Multi-Step Inequalities...80
Systems of Equations...81
Systems of Equations Word Problems...82

AFOQT Subject Test – Mathematics

Answers of Worksheets 83

Chapter 7 : Linear Functions 87
Finding Slope 88
Graphing Lines Using Line Equation 89
Writing Linear Equations 90
Graphing Linear Inequalities 91
Finding Midpoint 92
Finding Distance of Two Points 93
Answers of Worksheets 94

Chapter 8 : Polynomials 97
Writing Polynomials in Standard Form 98
Simplifying Polynomials 99
Adding and Subtracting Polynomials 100
Multiplying Monomials 101
Multiplying and Dividing Monomials 102
Multiplying a Polynomial and a Monomial 103
Multiplying Binomials 104
Factoring Trinomials 105
Operations with Polynomials 106
Answers of Worksheets 107

Chapter 9 : Functions Operations and Quadratic 111
Evaluating Function 112
Adding and Subtracting Functions 113
Multiplying and Dividing Functions 114
Composition of Functions 115
Quadratic Equation 116
Solving Quadratic Equations 117
Quadratic Formula and the Discriminant 118
Graphing Quadratic Functions 119
Answers of Worksheets 120

Chapter 10 : Geometry and Solid Figures 123
Angles 124
Pythagorean Relationship 125

WWW.MathNotion.Com

AFOQT Subject Test – Mathematics

Triangles 126

Polygons 127

Trapezoids 128

Circles 129

Cubes 130

Rectangular Prism 131

Cylinder 132

Pyramids and Cone 133

Answers of Worksheets 134

Chapter 11 : Statistics and Probability 137

Mean and Median 138

Mode and Range 139

Times Series 140

Stem–and–Leaf Plot 141

Pie Graph 142

Probability Problems 143

Answers of Worksheets 144

Chapter 12 : AFOQT Math Practice Tests 147

AFOQT Practice Test 1 149

   Arithmetic Reasoning 149

   Mathematics Knowledge 157

AFOQT Practice Test 2 163

   Arithmetic Reasoning 163

   Mathematics Knowledge 171

Chapter 13 : Answers and Explanations 177

Answer Key 177

Practice Test 1 179

   Arithmetic Reasoning 179

   Mathematics Knowledge 183

Practice Test 2 187

   Arithmetic Reasoning 187

   Mathematics Knowledge 191

AFOQT Subject Test – Mathematics

Chapter 1 :
Integers and Number Theory

Topics that you'll practice in this chapter:

- ✓ Rounding
- ✓ Whole Number Addition and Subtraction
- ✓ Whole Number Multiplication and Division
- ✓ Rounding and Estimates
- ✓ Adding and Subtracting Integers
- ✓ Multiplying and Dividing Integers
- ✓ Order of Operations
- ✓ Ordering Integers and Numbers
- ✓ Integers and Absolute Value
- ✓ Factoring Numbers
- ✓ Greatest Common Factor (GCF)
- ✓ Least Common Multiple (LCM)

"Wherever there is number, there is beauty." –Proclus

AFOQT Subject Test – Mathematics

Rounding

✍ **Round each number to the nearest ten.**

1) 42 = ____   5) 19 = ____   9) 48 = ____

2) 88 = ____   6) 25 = ____   10) 81 = ____

3) 24 = ____   7) 93 = ____   11) 58 = ____

4) 57 = ____   8) 71 = ____   12) 87 = ____

✍ **Round each number to the nearest hundred.**

13) 198 = ____   17) 321 = ____   21) 580 = ____

14) 387 = ____   18) 433 = ____   22) 868 = ____

15) 816 = ____   19) 579 = ____   23) 480 = ____

16) 101 = ____   20) 825 = ____   24) 287 = ____

✍ **Round each number to the nearest thousand.**

25) 1,382 = ____   29) 9,099 = ____   33) 52,866 = ____

26) 3,420 = ____   30) 22,980 = ____   34) 85,190 = ____

27) 4,254 = ____   31) 45,188 = ____   35) 70,990 = ____

28) 6,861 = ____   32) 16,808 = ____   36) 26,869 = ____

WWW.MathNotion.Com

AFOQT Subject Test – Mathematics

Whole Number Addition and Subtraction

✎ Find the sum or subtract.

1) 1,240 + 658 = _____

2) 3,458 − 544 = _____

3) 2,259 − 752 = _____

4) 1,990 + 324 = _____

5) 3,088 + 229 = _____

6) 2,354 + 1,009 = _____

7) 2,855 + 4,455 = _____

8) 5,112 + 4,004 = _____

9) 4,822 − 2,007 = _____

10) 8,380 − 5,288 = _____

11) 3,227 + 4,150 = _____

12) 7,702 − 4,331 = _____

✎ Find the missing number.

13) 720 + ____ = 1,360

14) 2,115 − ____ = 1,103

15) ____ + 3,105 = 6,200

16) 5,250 − 3,280 = ____

17) 8,020 + ____ = 8,990

18) 7,302 − 4,700 = ____

AFOQT Subject Test – Mathematics

Whole Number Multiplication and Division

✎ **Calculate each product.**

1) $\begin{array}{r} 42 \\ \times\ 13 \\ \hline \end{array}$

3) $\begin{array}{r} 40 \\ \times\ 14 \\ \hline \end{array}$

5) $\begin{array}{r} 110 \\ \times\ 11 \\ \hline \end{array}$

2) $\begin{array}{r} 70 \\ \times\ 15 \\ \hline \end{array}$

4) $\begin{array}{r} 22 \\ \times\ 20 \\ \hline \end{array}$

6) $\begin{array}{r} 150 \\ \times\ 16 \\ \hline \end{array}$

✎ **Find the missing quotient.**

7) $564 \div 6 = $ _____

8) $270 \div 3 = $ _____

9) $640 \div 8 = $ _____

10) $450 \div 9 = $ _____

11) $112 \div 7 = $ _____

12) $1{,}260 \div 9 = $ _____

13) $3{,}000 \div 10 = $ _____

14) $2{,}400 \div 8 = $ _____

15) $3{,}200 \div 40 = $ _____

16) $6{,}300 \div 90 = $ _____

✎ **Calculate each problem.**

17) $400 \div 5 = N$, $N = $ __

18) $2{,}500 \div 10 = N$, $N = $ __

19) $N \div 3 = 150$, $N = $ __

20) $42 \times N = 252$, $N = $ __

21) $660 \div N = 330$, $N = $ __

22) $N \times 8 = 336$, $N = $ __

AFOQT Subject Test – Mathematics

Rounding and Estimates

✎ **Estimate the sum by rounding each number to the nearest ten.**

1) $13 + 22 =$ _____

2) $71 + 23 =$ _____

3) $61 + 58 =$ _____

4) $56 + 85 =$ _____

5) $368 + 249 =$ _____

6) $330 + 903 =$ _____

7) $471 + 293 =$ _____

8) $1,950 + 2,655 =$ _____

✎ **Estimate the product by rounding each number to the nearest ten.**

9) $32 \times 71 =$ _____

10) $12 \times 33 =$ _____

11) $31 \times 83 =$ _____

12) $19 \times 11 =$ _____

13) $42 \times 76 =$ _____

14) $63 \times 34 =$ _____

15) $19 \times 31 =$ _____

16) $59 \times 71 =$ _____

✎ **Estimate the sum or product by rounding each number to the nearest ten.**

17) $\begin{array}{r} 29 \\ \times\ 12 \\ \hline \end{array}$

18) $\begin{array}{r} 37 \\ \times\ 26 \\ \hline \end{array}$

19) $\begin{array}{r} 48 \\ +\ 82 \\ \hline \end{array}$

20) $\begin{array}{r} 65 \\ +44 \\ \hline \end{array}$

21) $\begin{array}{r} 37 \\ \times\ 14 \\ \hline \end{array}$

22) $\begin{array}{r} 71 \\ +\ 32 \\ \hline \end{array}$

WWW.MathNotion.Com

AFOQT Subject Test – Mathematics

Adding and Subtracting Integers

✎ **Find each sum.**

1) $14 + (-6) =$

2) $(-13) + (-20) =$

3) $5 + (-28) =$

4) $50 + (-12) =$

5) $(-7) + (-15) + 3 =$

6) $30 + (-14) + 8 =$

7) $40 + (-10) + (-14) + 17 =$

8) $(-15) + (-20) + 13 + 35 =$

9) $40 + (-20) + (38 - 29) =$

10) $28 + (-12) + (30 - 12) =$

✎ **Find each difference.**

11) $(-18) - (-7) =$

12) $25 - (-14) =$

13) $(-20) - 36 =$

14) $34 - (-19) =$

15) $51 - (30 - 21) =$

16) $17 - (5) - (-24) =$

17) $(35 + 20) - (-46) =$

18) $48 - 16 - (-8) =$

19) $62 - (28 + 17) - (-15) =$

20) $58 - (-23) - (-31) =$

21) $19 - (-8) - (-13) =$

22) $(19 - 24) - (-14) =$

23) $27 - 33 - (-21) =$

24) $58 - (32 + 24) - (-9) =$

25) $36 - (-30) + (-17) =$

26) $27 - (-42) + (-31) =$

WWW.MathNotion.Com

Multiplying and Dividing Integers

✎ **Find each product.**

1) $(-9) \times (-5) =$

2) $(-3) \times 9 =$

3) $8 \times (-12) =$

4) $(-7) \times (-20) =$

5) $(-3) \times (-5) \times 6 =$

6) $(14 - 3) \times (-8) =$

7) $12 \times (-9) \times (-3) =$

8) $(140 + 10) \times (-2) =$

9) $10 \times (-12 + 8) \times 3 =$

10) $(-8) \times (-5) \times (-10) =$

✎ **Find each quotient.**

11) $42 \div (-7) =$

12) $(-48) \div (-6) =$

13) $(-40) \div (-8) =$

14) $54 \div (-2) =$

15) $152 \div 19 =$

16) $(-144) \div (-12) =$

17) $180 \div (-10) =$

18) $(-312) \div (-12) =$

19) $221 \div (-13) =$

20) $(-126) \div (6) =$

21) $(-161) \div (-7) =$

22) $-266 \div (-14) =$

23) $(-120) \div (-4) =$

24) $270 \div (-18) =$

25) $(-208) \div (-8) =$

26) $(135) \div (-15) =$

AFOQT Subject Test – Mathematics

Order of Operations

✎ **Evaluate each expression.**

1) $7 + (5 \times 4) =$

2) $14 - (3 \times 6) =$

3) $(19 \times 4) + 16 =$

4) $(16 - 7) - (8 \times 2) =$

5) $27 + (18 \div 3) =$

6) $(18 \times 8) \div 6 =$

7) $(32 \div 4) \times (-2) =$

8) $(9 \times 4) + (32 - 18) =$

9) $24 + (4 \times 3) + 7 =$

10) $(36 \times 3) \div (2 + 2) =$

11) $(-7) + (12 \times 3) + 11 =$

12) $(8 \times 5) - (24 \div 6) =$

13) $(7 \times 6 \div 3) - (12 + 9) =$

14) $(13 + 5 - 14) \times 3 - 2 =$

15) $(20 - 14 + 30) \times (64 \div 4) =$

16) $32 + (28 - (36 \div 9)) =$

17) $(7 + 6 - 4 - 7) + (15 \div 5) =$

18) $(85 - 20) + (20 - 18 + 7) =$

19) $(20 \times 2) + (14 \times 3) - 22 =$

20) $18 + 5 - (30 \times 3) + 20 =$

WWW.MathNotion.Com

Ordering Integers and Numbers

✎ **Order each set of integers from least to greatest.**

1) $8, -10, -5, -3, 4$ ___, ___, ___, ___, ___, ___

2) $-10, -18, 6, 14, 27$ ___, ___, ___, ___, ___, ___

3) $15, -8, -21, 21, -23$ ___, ___, ___, ___, ___, ___

4) $-14, -40, 23, -12, 47$ ___, ___, ___, ___, ___, ___

5) $59, -54, 32, -57, 36$ ___, ___, ___, ___, ___, ___

6) $68, 26, -19, 47, -34$ ___, ___, ___, ___, ___, ___

✎ **Order each set of integers from greatest to least.**

7) $18, 36, -16, -18, -10$ ___, ___, ___, ___, ___, ___

8) $27, 34, -12, -24, 94$ ___, ___, ___, ___, ___, ___

9) $50, -21, -13, 42, -2$ ___, ___, ___, ___, ___, ___

10) $37, 46, -20, -16, 86$ ___, ___, ___, ___, ___, ___

11) $-18, 88, -26, -59, 75$ ___, ___, ___, ___, ___, ___

12) $-65, -30, -25, 3, 14$ ___, ___, ___, ___, ___, ___

AFOQT Subject Test – Mathematics

Integers and Absolute Value

✎ Write absolute value of each number.

1) $|-2| =$

2) $|-27| =$

3) $|-20| =$

4) $|14| =$

5) $|6| =$

6) $|-55| =$

7) $|16| =$

8) $|2| =$

9) $|54| =$

10) $|-4| =$

11) $|-11|$

12) $|88| =$

13) $|0| =$

14) $|79| =$

15) $|-32| =$

16) $|-17| =$

17) $|42| =$

18) $|-46| =$

19) $|1| =$

20) $|-40| =$

✎ Evaluate the value.

21) $|-5| - \frac{|-21|}{7} =$

22) $14 - |3 - 15| - |-4| =$

23) $\frac{|-32|}{4} \times |-4| =$

24) $\frac{|7 \times (-3)|}{7} \times \frac{|-19|}{3} =$

25) $|4 \times (-5)| + \frac{|-40|}{5} =$

26) $\frac{|-45|}{9} \times \frac{|-24|}{12} =$

27) $|-12 + 8| \times \frac{|-7 \times 7|}{7} =$

28) $\frac{|-11 \times 2|}{4} \times |-16| =$

WWW.MathNotion.Com

AFOQT Subject Test – Mathematics

Factoring Numbers

✎ List all positive factors of each number.

1) 9

2) 16

3) 24

4) 30

5) 26

6) 46

7) 20

8) 68

9) 28

10) 98

11) 14

12) 54

13) 55

14) 18

15) 63

16) 34

17) 50

18) 62

19) 95

20) 64

21) 70

22) 45

23) 22

24) 65

AFOQT Subject Test – Mathematics

Greatest Common Factor

✏ **Find the GCF for each number pair.**

1) 6, 2

2) 4, 5

3) 3, 12

4) 7, 3

5) 5, 10

6) 8, 48

7) 6, 18

8) 9, 15

9) 12, 18

10) 4, 36

11) 6, 10

12) 28, 52

13) 25, 10

14) 22, 24

15) 9, 54

16) 8, 54

17) 42, 14

18) 16, 40

19) 9, 2, 3

20) 5, 15, 10

21) 7, 9, 2

22) 16, 64

23) 30, 48

24) 36, 63

WWW.MathNotion.Com

AFOQT Subject Test – Mathematics

Least Common Multiple

✏ **Find the LCM for each number pair.**

1) 6, 9

2) 15, 45

3) 16, 40

4) 12, 36

5) 18, 27

6) 14, 42

7) 6, 30

8) 8, 56

9) 7, 21

10) 8, 20

11) 15, 25

12) 7, 9

13) 4, 11

14) 8, 28

15) 28, 56

16) 40, 50

17) 12, 13

18) 22, 11

19) 36, 20

20) 15, 35

21) 18, 81

22) 30, 54

23) 18, 45

24) 75, 25

AFOQT Subject Test – Mathematics

Answers of Worksheets

Rounding

1) 40
2) 90
3) 20
4) 60
5) 20
6) 30
7) 90
8) 70
9) 50
10) 80
11) 60
12) 90
13) 200
14) 400
15) 800
16) 100
17) 300
18) 400
19) 600
20) 800
21) 600
22) 900
23) 500
24) 300
25) 1,000
26) 3,000
27) 4,000
28) 7,000
29) 9,000
30) 23,000
31) 45,000
32) 17,000
33) 53,000
34) 85,000
35) 71,000
36) 27,000

Whole Number Addition and Subtraction

1) 1,898
2) 2,914
3) 1,507
4) 2,314
5) 3,317
6) 3,363
7) 7,310
8) 9,116
9) 2,815
10) 3,092
11) 7,377
12) 3,371
13) 640
14) 1,012
15) 3,095
16) 1,970
17) 970
18) 2,602

Whole Number Multiplication and Division

1) 546
2) 1,050
3) 560
4) 440
5) 1,210
6) 2,400
7) 94
8) 90
9) 80
10) 50
11) 16
12) 140
13) 300
14) 300
15) 80
16) 70
17) 80
18) 250
19) 450
20) 6
21) 2
22) 42

Rounding and Estimates

1) 30
2) 90
3) 120
4) 150
5) 620
6) 1,230
7) 760
8) 4,610
9) 2,100
10) 300
11) 2,400
12) 200
13) 3,200
14) 1,800
15) 600
16) 4,200
17) 300
18) 1,200
19) 130
20) 110
21) 400
22) 100

AFOQT Subject Test – Mathematics

Adding and Subtracting Integers

1) 8
2) −33
3) −23
4) 38
5) −19
6) 24
7) 33
8) 13
9) 29
10) 34
11) −11
12) 39
13) −56
14) 53
15) 42
16) 36
17) 101
18) 40
19) 32
20) 112
21) 40
22) 9
23) 15
24) 11
25) 49
26) 38

Multiplying and Dividing Integers

1) 45
2) −27
3) −96
4) 140
5) 90
6) −88
7) 324
8) −300
9) −120
10) −400
11) −6
12) 8
13) 5
14) −27
15) 8
16) 12
17) −18
18) 26
19) −17
20) −21
21) 23
22) 19
23) 30
24) −15
25) 26
26) −9

Order of Operations

1) 27
2) −4
3) 92
4) −7
5) 33
6) 24
7) −16
8) 50
9) 43
10) 27
11) 40
12) 36
13) −7
14) 10
15) 576
16) 56
17) 5
18) 74
19) 60
20) −47

Ordering Integers and Numbers

1) −10, −5, −3, 4, 8
2) −18, −10, 6, 14, 27
3) −23, −21, −8, 15, 21
4) −40, −14, −12, 23, 47
5) −57, −54, 32, 36, 59
6) −34, −19, 26, 47, 68
7) 36, 18, −10, −16, −18
8) 94, 34, 27, −12, −24
9) 50, 42, −2, −13, −21
10) 86, 46, 37, −16, −20
11) 88, 75, −18, −26, −59
12) 14, 3, −25, −30, −65

Integers and Absolute Value

1) 2
2) 27
3) 20
4) 14

WWW.MathNotion.Com

AFOQT Subject Test – Mathematics

5) 6	11) 11	17) 42	23) 32
6) 55	12) 88	18) 46	24) 19
7) 16	13) 0	19) 1	25) 28
8) 2	14) 79	20) 40	26) 10
9) 54	15) 32	21) 2	27) 28
10) 4	16) 17	22) −2	28) 88

Factoring Numbers

1) 1, 3, 9	10) 1, 2, 7, 14, 49, 98	19) 1, 5, 19, 95
2) 1, 2, 4, 8, 16	11) 1, 2, 7, 14	20) 1, 2, 4, 8, 16, 32, 64
3) 1, 2, 3, 4, 6, 8, 12, 24	12) 1, 2, 3, 6, 9, 18, 27, 54	21) 1, 2, 5, 7, 10, 14, 35, 70
4) 1, 2, 3, 5, 6, 10, 15, 30	13) 1, 5, 11, 55	22) 1, 3, 5, 9, 15, 45
5) 1, 2, 13, 26	14) 1, 2, 3, 6, 9, 18	23) 1, 2, 11, 22
6) 1, 2, 23, 46	15) 1, 3, 7, 9, 21, 63	24) 1, 5, 13, 65
7) 1, 2, 4, 5, 10, 20	16) 1, 2, 17, 34	
8) 1, 2, 4, 17, 34, 68	17) 1, 2, 5, 10, 25, 50	
9) 1, 2, 4, 7, 14, 28	18) 1, 2, 31, 62	

Greatest Common Factor

1) 2	7) 6	13) 5	19) 1
2) 1	8) 3	14) 2	20) 5
3) 3	9) 6	15) 9	21) 1
4) 1	10) 4	16) 2	22) 16
5) 5	11) 2	17) 14	23) 6
6) 8	12) 4	18) 8	24) 9

Least Common Multiple

1) 18	7) 30	13) 44	19) 180
2) 45	8) 56	14) 56	20) 105
3) 80	9) 21	15) 56	21) 162
4) 36	10) 40	16) 200	22) 270
5) 54	11) 75	17) 156	23) 90
6) 42	12) 63	18) 22	24) 75

AFOQT Subject Test – Mathematics

Chapter 2 :
Fractions and Decimals

Topics that you'll practice in this chapter:

- ✓ Simplifying Fractions
- ✓ Adding and Subtracting Fractions
- ✓ Multiplying and Dividing Fractions
- ✓ Adding and Subtract Mixed Numbers
- ✓ Multiplying and Dividing Mixed Numbers
- ✓ Adding and Subtracting Decimals
- ✓ Multiplying and Dividing Decimals
- ✓ Comparing Decimals
- ✓ Rounding Decimals

"A Man is like a fraction whose numerator is what he is and whose denominator is what he thinks of himself. The larger the denominator, the smaller the fraction." –Tolstoy

AFOQT Subject Test – Mathematics

Simplifying Fractions

✎ **Simplify each fraction to its lowest terms.**

1) $\frac{5}{10} =$

2) $\frac{28}{35} =$

3) $\frac{27}{36} =$

4) $\frac{40}{80} =$

5) $\frac{14}{56} =$

6) $\frac{32}{48} =$

7) $\frac{52}{65} =$

8) $\frac{15}{60} =$

9) $\frac{80}{160} =$

10) $\frac{55}{77} =$

11) $\frac{28}{112} =$

12) $\frac{32}{64} =$

13) $\frac{63}{72} =$

14) $\frac{81}{90} =$

15) $\frac{35}{105} =$

16) $\frac{25}{70} =$

17) $\frac{80}{280} =$

18) $\frac{12}{81} =$

19) $\frac{36}{186} =$

20) $\frac{240}{540} =$

21) $\frac{70}{560} =$

✎ **Find the answer for each problem.**

22) Which of the following fractions equal to $\frac{3}{4}$? ____

    A. $\frac{60}{90}$ B. $\frac{43}{104}$ C. $\frac{48}{64}$ D. $\frac{150}{300}$

23) Which of the following fractions equal to $\frac{5}{8}$? ____

    A. $\frac{125}{200}$ B. $\frac{115}{200}$ C. $\frac{50}{100}$ D. $\frac{30}{90}$

24) Which of the following fractions equal to $\frac{3}{7}$? ____

    A. $\frac{58}{116}$ B. $\frac{54}{126}$ C. $\frac{270}{167}$ D. $\frac{42}{63}$

AFOQT Subject Test – Mathematics

Adding and Subtracting Fractions

✏️ **Find the sum.**

1) $\frac{5}{9} + \frac{4}{9} =$

2) $\frac{1}{2} + \frac{1}{7} =$

3) $\frac{3}{8} + \frac{1}{4} =$

4) $\frac{3}{5} + \frac{1}{2} =$

5) $\frac{1}{4} + \frac{3}{5} =$

6) $\frac{7}{8} + \frac{3}{8} =$

7) $\frac{1}{2} + \frac{7}{10} =$

8) $\frac{2}{5} + \frac{2}{3} =$

9) $\frac{5}{7} + \frac{2}{3} =$

10) $\frac{7}{12} + \frac{3}{4} =$

11) $\frac{5}{6} + \frac{2}{5} =$

12) $\frac{1}{12} + \frac{2}{3} =$

✏️ **Find the difference.**

13) $\frac{1}{3} - \frac{1}{6} =$

14) $\frac{3}{4} - \frac{1}{8} =$

15) $\frac{1}{2} - \frac{1}{3} =$

16) $\frac{1}{4} - \frac{1}{5} =$

17) $\frac{5}{8} - \frac{2}{3} =$

18) $\frac{1}{4} - \frac{1}{7} =$

19) $\frac{5}{6} - \frac{1}{9} =$

20) $\frac{3}{4} - \frac{1}{6} =$

21) $\frac{7}{8} - \frac{1}{12} =$

22) $\frac{8}{15} - \frac{3}{5} =$

23) $\frac{3}{12} - \frac{1}{14} =$

24) $\frac{10}{13} - \frac{7}{26} =$

25) $\frac{6}{7} - \frac{3}{4} =$

26) $\frac{4}{5} - \frac{1}{8} =$

27) $\frac{4}{7} - \frac{2}{35} =$

28) $\frac{9}{16} - \frac{2}{8} =$

29) $\frac{8}{9} - \frac{7}{18} =$

30) $\frac{1}{2} - \frac{4}{9} =$

WWW.MathNotion.Com

AFOQT Subject Test – Mathematics

Multiplying and Dividing Fractions

✎ Find the value of each expression in lowest terms.

1) $\dfrac{1}{5} \times \dfrac{15}{5} =$

2) $\dfrac{9}{12} \times \dfrac{4}{9} =$

3) $\dfrac{1}{16} \times \dfrac{8}{10} =$

4) $\dfrac{1}{24} \times \dfrac{8}{10} =$

5) $\dfrac{1}{5} \times \dfrac{1}{4} =$

6) $\dfrac{7}{9} \times \dfrac{1}{7} =$

7) $\dfrac{6}{7} \times \dfrac{1}{3} =$

8) $\dfrac{2}{8} \times \dfrac{2}{8} =$

9) $\dfrac{5}{8} \times \dfrac{3}{5} =$

10) $\dfrac{4}{7} \times \dfrac{1}{8} =$

11) $\dfrac{7}{15} \times \dfrac{5}{7} =$

12) $\dfrac{3}{10} \times \dfrac{5}{9} =$

✎ Find the value of each expression in lowest terms.

13) $\dfrac{1}{4} \div \dfrac{1}{8} =$

14) $\dfrac{1}{10} \div \dfrac{1}{5} =$

15) $\dfrac{3}{4} \div \dfrac{1}{5} =$

16) $\dfrac{1}{3} \div \dfrac{5}{6} =$

17) $\dfrac{1}{7} \div \dfrac{8}{42} =$

18) $\dfrac{3}{4} \div \dfrac{1}{6} =$

19) $\dfrac{2}{7} \div \dfrac{7}{13} =$

20) $\dfrac{1}{24} \div \dfrac{3}{16} =$

21) $\dfrac{7}{12} \div \dfrac{5}{6} =$

22) $\dfrac{22}{18} \div \dfrac{11}{9} =$

23) $\dfrac{9}{35} \div \dfrac{3}{7} =$

24) $\dfrac{2}{7} \div \dfrac{8}{21} =$

25) $\dfrac{1}{9} \div \dfrac{2}{5} =$

26) $\dfrac{5}{12} \div \dfrac{3}{5} =$

27) $\dfrac{3}{20} \div \dfrac{1}{6} =$

28) $\dfrac{8}{20} \div \dfrac{3}{4} =$

29) $\dfrac{5}{6} \div \dfrac{2}{9} =$

30) $\dfrac{5}{11} \div \dfrac{3}{4} =$

AFOQT Subject Test – Mathematics

Adding and Subtracting Mixed Numbers

✏️ **Find the sum.**

1) $3\frac{1}{3} + 2\frac{1}{6} =$

2) $4\frac{1}{2} + 3\frac{1}{2} =$

3) $3\frac{3}{8} + 1\frac{1}{8} =$

4) $2\frac{1}{4} + 2\frac{1}{3} =$

5) $3\frac{5}{6} + 2\frac{7}{12} =$

6) $5\frac{4}{15} + 3\frac{3}{5} =$

7) $2\frac{1}{3} + 4\frac{3}{7} =$

8) $3\frac{1}{2} + 4\frac{2}{5} =$

9) $5\frac{2}{5} + 6\frac{3}{7} =$

10) $8\frac{5}{16} + 6\frac{1}{12} =$

✏️ **Find the difference.**

11) $3\frac{1}{4} - 1\frac{3}{4} =$

12) $6\frac{3}{5} - 4\frac{2}{5} =$

13) $4\frac{1}{3} - 3\frac{1}{9} =$

14) $7\frac{1}{7} - 5\frac{1}{2} =$

15) $5\frac{1}{3} - 2\frac{1}{12} =$

16) $8\frac{1}{5} - 4\frac{1}{3} =$

17) $9\frac{1}{4} - 6\frac{1}{8} =$

18) $11\frac{7}{15} - 8\frac{3}{5} =$

19) $14\frac{5}{6} - 11\frac{3}{5} =$

20) $18\frac{2}{7} - 14\frac{1}{5} =$

21) $9\frac{1}{3} - 4\frac{1}{4} =$

22) $6\frac{1}{8} - 4\frac{1}{16} =$

23) $19\frac{3}{8} - 15\frac{1}{3} =$

24) $11\frac{1}{9} - 8\frac{1}{8} =$

25) $17\frac{1}{7} - 11\frac{1}{5} =$

26) $16\frac{2}{9} - 9\frac{5}{7} =$

WWW.MathNotion.Com

AFOQT Subject Test – Mathematics

Multiplying and Dividing Mixed Numbers

✎ **Find the product.**

1) $5\frac{1}{2} \times 2\frac{1}{4} =$

2) $5\frac{1}{3} \times 4\frac{1}{3} =$

3) $5\frac{3}{4} \times 6\frac{1}{4} =$

4) $3\frac{1}{3} \times 2\frac{3}{5} =$

5) $4\frac{8}{10} \times 1\frac{1}{24} =$

6) $6\frac{2}{7} \times 1\frac{1}{11} =$

7) $8\frac{2}{3} \times 3\frac{1}{2} =$

8) $3\frac{4}{7} \times 2\frac{1}{5} =$

9) $5\frac{2}{8} \times 4\frac{1}{6} =$

10) $7\frac{3}{3} \times 1\frac{3}{8} =$

✎ **Find the quotient.**

11) $2\frac{2}{5} \div 4\frac{1}{5} =$

12) $4\frac{1}{6} \div 3\frac{1}{3} =$

13) $6\frac{1}{3} \div 1\frac{1}{2} =$

14) $7\frac{1}{10} \div 2\frac{2}{5} =$

15) $3\frac{1}{3} \div 1\frac{1}{9} =$

16) $1\frac{1}{10} \div 4\frac{1}{2} =$

17) $1\frac{3}{16} \div 5\frac{1}{4} =$

18) $4\frac{1}{3} \div 4\frac{3}{4} =$

19) $9\frac{1}{3} \div 2\frac{1}{4} =$

20) $15\frac{1}{3} \div 5\frac{1}{2} =$

21) $4\frac{1}{6} \div 1\frac{1}{5} =$

22) $1\frac{1}{18} \div 1\frac{2}{9} =$

23) $4\frac{2}{7} \div 1\frac{3}{10} =$

24) $7\frac{1}{3} \div 2\frac{2}{11} =$

25) $8\frac{2}{5} \div 1\frac{1}{6} =$

26) $9\frac{1}{3} \div 2\frac{1}{7} =$

WWW.MathNotion.Com

AFOQT Subject Test – Mathematics

Adding and Subtracting Decimals

✍ **Add and subtract decimals.**

1) 35.19 − 24.28 = ____

4) 38.72 − 21.68 = ____

7) 86.09 − 35.14 = ____

2) 34.29 + 42.58 = ____

5) 57.39 + 26.54 = ____

8) 54.51 + 32.66 = ____

3) 61.20 + 33.75 = ____

6) 70.24 − 42.35 = ____

9) 114.21 − 88.69 = ____

✍ **Find the missing number.**

10) ___ + 2.8 = 5.4

11) 4.1 + ___ = 5.88

12) 6.45 + ___ = 8

13) 7.25 − ___ = 3.40

14) ___ − 2.35 = 4.25

15) ___ − 19.85 = 6.54

16) 22.15 + ___ = 28.95

17) ___ − 37.16 = 9.42

18) ___ + 24.50 = 34.19

19) 72.40 + ___ = 125.20

WWW.MathNotion.Com

> AFOQT Subject Test – Mathematics

Multiplying and Dividing Decimals

✎ **Find the product.**

1) 0.5 × 0.6 =

2) 3.3 × 0.4 =

3) 1.28 × 0.5 =

4) 0.35 × 0.6 =

5) 1.85 × 0.6 =

6) 0.24 × 0.5 =

7) 5.25 × 1.4 =

8) 18.5 × 4.6 =

9) 15.4 × 6.8 =

10) 19.5 × 2.6 =

11) 32.2 × 1.5 =

12) 78.4 × 4.5 =

✎ **Find the quotient.**

13) 1.85 ÷ 10 =

14) 74.6 ÷ 100 =

15) 3.6 ÷ 3 =

16) 9.6 ÷ 0.4 =

17) 15.5 ÷ 0.5 =

18) 32.8 ÷ 0.2 =

19) 22.15 ÷ 1,000 =

20) 53.55 ÷ 0.7 =

21) 322.2 ÷ 0.2 =

22) 50.67 ÷ 0.18 =

23) 77.4 ÷ 0.8 =

24) 27.93 ÷ 0.03 =

Comparing Decimals

✎ Write the correct comparison symbol (>, < or =).

1) 0.70 ☐ 0.070

2) 0.049 ☐ 0.49

3) 5.090 ☐ 5.09

4) 2.57 ☐ 2.05

5) 9.03 ☐ 0.930

6) 6.06 ☐ 6.6

7) 7.02 ☐ 7.020

8) 3.04 ☐ 3.2

9) 3.61 ☐ 3.245

10) 0.986 ☐ 0.0986

11) 17.24 ☐ 17.240

12) 0.759 ☐ 0.81

13) 9.040 ☐ 9.40

14) 5.73 ☐ 5.213

15) 9.44 ☐ 9.404

16) 7.17 ☐ 7.170

17) 4.85 ☐ 4.085

18) 9.041 ☐ 9.40

19) 3.033 ☐ 3.030

20) 4.97 ☐ 4.970

AFOQT Subject Test – Mathematics

Rounding Decimals

✍ **Round each decimal to the nearest whole number.**

1) 28.12   3) 16.22   5) 7.95

2) 6.9   4) 8.5   6) 52.7

✍ **Round each decimal to the nearest tenth.**

7) 31.761   9) 94.729   11) 13.219

8) 14.421   10) 77.89   12) 59.89

✍ **Round each decimal to the nearest hundredth.**

13) 8.428   15) 55.3786   17) 62.241

14) 23.812   16) 231.912   18) 19.447

✍ **Round each decimal to the nearest thousandth.**

19) 15.54324   21) 243.8652   23) 67.1983

20) 34.62586   22) 80.4529   24) 72.36788

AFOQT Subject Test – Mathematics

Answers of Worksheets

Simplifying Fractions

1) $\frac{1}{2}$
2) $\frac{4}{5}$
3) $\frac{3}{4}$
4) $\frac{1}{2}$
5) $\frac{1}{4}$
6) $\frac{2}{3}$
7) $\frac{4}{5}$
8) $\frac{1}{4}$
9) $\frac{1}{2}$
10) $\frac{5}{7}$
11) $\frac{1}{4}$
12) $\frac{1}{2}$
13) $\frac{7}{8}$
14) $\frac{9}{10}$
15) $\frac{1}{3}$
16) $\frac{5}{14}$
17) $\frac{2}{7}$
18) $\frac{4}{27}$
19) $\frac{6}{31}$
20) $\frac{4}{9}$
21) $\frac{1}{8}$
22) C
23) A
24) B

Adding and Subtracting Fractions

1) $\frac{9}{9} = 1$
2) $\frac{9}{14}$
3) $\frac{5}{8}$
4) $1\frac{1}{10}$
5) $\frac{17}{20}$
6) $1\frac{1}{4}$
7) $1\frac{1}{5}$
8) $1\frac{1}{15}$
9) $1\frac{8}{21}$
10) $1\frac{1}{3}$
11) $1\frac{7}{30}$
12) $\frac{3}{4}$
13) $\frac{1}{6}$
14) $\frac{5}{8}$
15) $\frac{1}{6}$
16) $\frac{1}{20}$
17) $-\frac{1}{24}$
18) $\frac{3}{28}$
19) $\frac{13}{18}$
20) $\frac{7}{12}$
21) $\frac{19}{24}$
22) $-\frac{1}{15}$
23) $\frac{5}{28}$
24) $\frac{1}{2}$
25) $\frac{3}{28}$
26) $\frac{27}{40}$
27) $\frac{18}{35}$
28) $\frac{5}{16}$
29) $\frac{1}{2}$
30) $\frac{1}{18}$

Multiplying and Dividing Fractions

1) $\frac{3}{5}$
2) $\frac{1}{3}$
3) $\frac{1}{20}$
4) $\frac{1}{30}$
5) $\frac{1}{20}$
6) $\frac{1}{9}$
7) $\frac{2}{7}$
8) $\frac{1}{16}$
9) $\frac{3}{8}$
10) $\frac{1}{14}$
11) $\frac{1}{3}$
12) $\frac{1}{6}$
13) 2
14) $\frac{1}{2}$

WWW.MathNotion.Com

AFOQT Subject Test – Mathematics

15) $3\frac{3}{4}$ 19) $\frac{26}{49}$ 23) $\frac{3}{5}$ 27) $\frac{9}{10}$

16) $\frac{2}{5}$ 20) $\frac{2}{9}$ 24) $\frac{3}{4}$ 28) $\frac{8}{15}$

17) $\frac{3}{4}$ 21) $\frac{7}{10}$ 25) $\frac{5}{18}$ 29) $3\frac{3}{4}$

18) $4\frac{1}{2}$ 22) 1 26) $\frac{25}{36}$ 30) $\frac{20}{33}$

Adding and Subtracting Mixed Numbers

1) $5\frac{1}{2}$ 8) $7\frac{9}{10}$ 15) $3\frac{1}{4}$ 22) $2\frac{1}{16}$

2) 8 9) $11\frac{29}{35}$ 16) $3\frac{13}{15}$ 23) $4\frac{1}{24}$

3) $4\frac{1}{2}$ 10) $14\frac{19}{48}$ 17) $3\frac{1}{8}$ 24) $2\frac{71}{72}$

4) $4\frac{7}{12}$ 11) $1\frac{1}{2}$ 18) $2\frac{13}{15}$ 25) $5\frac{33}{35}$

5) $6\frac{5}{12}$ 12) $2\frac{1}{5}$ 19) $3\frac{7}{30}$ 26) $6\frac{32}{63}$

6) $8\frac{13}{15}$ 13) $1\frac{2}{9}$ 20) $4\frac{3}{35}$

7) $6\frac{16}{21}$ 14) $1\frac{9}{14}$ 21) $5\frac{1}{12}$

Multiplying and Dividing Mixed Numbers

1) $12\frac{3}{8}$ 10) 11 19) $4\frac{4}{27}$

2) $23\frac{1}{9}$ 11) $\frac{4}{7}$ 20) $2\frac{26}{33}$

3) $35\frac{15}{16}$ 12) $1\frac{1}{4}$ 21) $3\frac{17}{36}$

4) $8\frac{2}{3}$ 13) $4\frac{2}{9}$ 22) $\frac{19}{22}$

5) 5 14) $2\frac{23}{24}$ 23) $3\frac{27}{91}$

6) $6\frac{6}{7}$ 15) 3 24) $3\frac{13}{36}$

7) $30\frac{1}{3}$ 16) $\frac{11}{45}$ 25) $7\frac{1}{5}$

8) $7\frac{6}{7}$ 17) $\frac{19}{84}$ 26) $4\frac{16}{45}$

9) $21\frac{7}{8}$ 18) $\frac{52}{57}$

Adding and Subtracting Decimals

1) 10.91   2) 76.87   3) 94.95   4) 17.04

AFOQT Subject Test – Mathematics

5) 83.93	9) 25.52	13) 3.85	17) 46.58
6) 27.89	10) 2.6	14) 6.6	18) 9.69
7) 50.95	11) 1.78	15) 26.39	19) 52.8
8) 87.17	12) 1.55	16) 6.8	

Multiplying and Dividing Decimals

1) 0.3	7) 7.35	13) 0.185	19) 0.02215
2) 1.32	8) 85.1	14) 0.746	20) 76.5
3) 0.64	9) 104.72	15) 1.2	21) 1,611
4) 0.21	10) 50.7	16) 24	22) 281.5
5) 1.11	11) 48.3	17) 31	23) 96.75
6) 0.12	12) 352.8	18) 164	24) 931

Comparing Decimals

1) >	6) <	11) =	16) =
2) <	7) =	12) <	17) >
3) =	8) <	13) <	18) <
4) >	9) >	14) >	19) >
5) >	10) >	15) >	20) =

Rounding Decimals

1) 28	9) 94.7	17) 62.24
2) 7	10) 77.9	18) 19.45
3) 16	11) 13.2	19) 15.543
4) 9	12) 59.9	20) 34.626
5) 8	13) 8.43	21) 243.865
6) 53	14) 23.81	22) 80.453
7) 31.8	15) 55.38	23) 67.198
8) 14.4	16) 231.91	24) 72.368

WWW.MathNotion.Com

AFOQT Subject Test – Mathematics

Chapter 3 : Proportions, Ratios, and Percent

Topics that you'll practice in this chapter:

- ✓ Simplifying Ratios
- ✓ Proportional Ratios
- ✓ Similarity and Ratios
- ✓ Ratio and Rates Word Problems
- ✓ Percentage Calculations
- ✓ Percent Problems
- ✓ Discount, Tax and Tip
- ✓ Percent of Change
- ✓ Simple Interest

Without mathematics, there's nothing you can do. Everything around you is mathematics. Everything around you is numbers." – Shakuntala Devi

AFOQT Subject Test – Mathematics

Simplifying Ratios

✍ Reduce each ratio.

1) $15:20 =$ ___ : ___

2) $7:70 =$ ___ : ___

3) $16:28 =$ ___ : ___

4) $7:21 =$ ___ : ___

5) $4:40 =$ ___ : ___

6) $6:48 =$ ___ : ___

7) $16:64 =$ ___ : ___

8) $10:25 =$ ___ : ___

9) $8:48 =$ ___ : ___

10) $49:63 =$ ___ : ___

11) $18:27 =$ ___ : ___

12) $35:10 =$ ___ : ___

13) $90:9 =$ ___ : ___

14) $24:32 =$ ___ : ___

15) $7:56 =$ ___ : ___

16) $45:63 =$ ___ : ___

17) $56:72 =$ ___ : ___

18) $26:13 =$ ___ : ___

19) $15:45 =$ ___ : ___

20) $28:4 =$ ___ : ___

21) $24:48 =$ ___ : ___

22) $30:24 =$ ___ : ___

23) $70:140 =$ ___ : ___

24) $6:180 =$ ___ : ___

✍ Write each ratio as a fraction in simplest form.

25) $6:12 =$

26) $30:50 =$

27) $15:35 =$

28) $9:27 =$

29) $8:24 =$

30) $18:84 =$

31) $7:14 =$

32) $7:35 =$

33) $40:96 =$

34) $12:54 =$

35) $44:52 =$

36) $12:27 =$

37) $15:180 =$

38) $39:143 =$

39) $20:300 =$

40) $30:120 =$

41) $56:42 =$

42) $26:130 =$

43) $66:123 =$

44) $70:630 =$

45) $75:125 =$

AFOQT Subject Test – Mathematics

Proportional Ratios

✎ **Fill in the blanks; Calculate each proportion.**

1) $3:8 = \_\_ : 48$

2) $2:5 = 20:\_\_$

3) $1:9 = \_\_ : 81$

4) $6:7 = 12:\_\_$

5) $9:2 = 63:\_\_$

6) $8:7 = \_\_ : 49$

7) $20:3 = \_\_ : 15$

8) $1:3 = \_\_ : 75$

9) $7:6 = \_\_ : 60$

10) $8:5 = \_\_ : 45$

11) $3:10 = 60:\_\_$

12) $6:11 = 42:\_\_$

✎ **State if each pair of ratios form a proportion.**

13) $\frac{3}{20}$ and $\frac{9}{60}$

14) $\frac{1}{7}$ and $\frac{6}{42}$

15) $\frac{3}{7}$ and $\frac{24}{56}$

16) $\frac{4}{9}$ and $\frac{12}{18}$

17) $\frac{1}{9}$ and $\frac{12}{81}$

18) $\frac{7}{8}$ and $\frac{21}{28}$

19) $\frac{9}{13}$ and $\frac{27}{39}$

20) $\frac{1}{8}$ and $\frac{8}{64}$

21) $\frac{6}{19}$ and $\frac{30}{85}$

22) $\frac{5}{9}$ and $\frac{40}{81}$

23) $\frac{9}{14}$ and $\frac{108}{168}$

24) $\frac{15}{23}$ and $\frac{360}{552}$

✎ **Calculate each proportion.**

25) $\frac{20}{25} = \frac{32}{x}, x = \_\_\_$

26) $\frac{1}{8} = \frac{32}{x}, x = \_\_\_$

27) $\frac{15}{5} = \frac{21}{x}, x = \_\_\_$

28) $\frac{1}{7} = \frac{x}{294}, x = \_\_\_$

29) $\frac{7}{9} = \frac{x}{81}, x = \_\_\_$

30) $\frac{1}{5} = \frac{13}{x}, x = \_\_\_$

31) $\frac{9}{5} = \frac{36}{x}, x = \_\_\_$

32) $\frac{6}{13} = \frac{48}{x}, x = \_\_\_$

33) $\frac{5}{8} = \frac{x}{88}, x = \_\_\_$

34) $\frac{4}{15} = \frac{x}{240}, x = \_\_\_$

35) $\frac{9}{19} = \frac{x}{266}, x = \_\_\_$

36) $\frac{7}{15} = \frac{x}{270}, x = \_\_\_$

AFOQT Subject Test – Mathematics

Similarity and Ratios

✎ **Each pair of figures is similar. Find the missing side.**

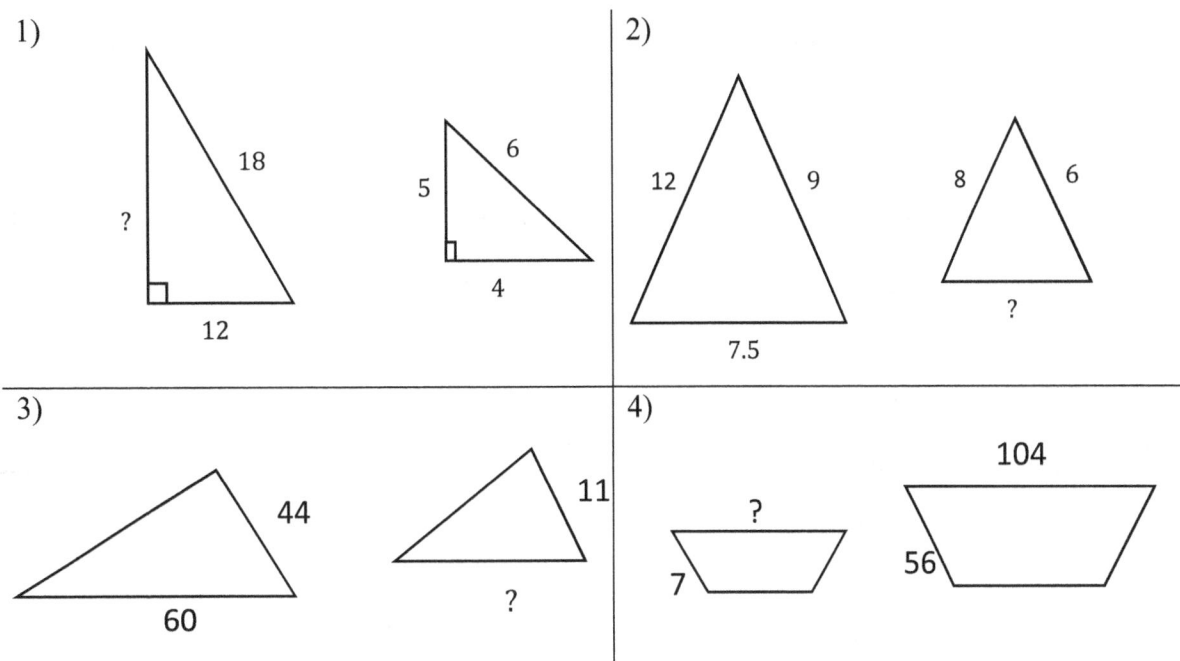

✎ **Calculate.**

5) Two rectangles are similar. The first is 24 feet wide and 120 feet long. The second is 30 feet wide. What is the length of the second rectangle? _____

6) Two rectangles are similar. One is 5 meters by 36 meters. The longer side of the second rectangle is 90 meters. What is the other side of the second rectangle? _____

7) A building casts a shadow 25 ft long. At the same time a girl 10 ft tall casts a shadow 5 ft long. How tall is the building? _____

8) The scale of a map of Texas is 4 inches: 32 miles. If you measure the distance from Dallas to Martin County as 38.4 inches, approximately how far is Martin County from Dallas? _____

AFOQT Subject Test – Mathematics

Ratio and Rates Word Problems

✎ **Find the answer for each word problem.**

1) Mason has 24 red cards and 36 green cards. What is the ratio of Mason's red cards to his green cards? _____

2) In a party, 45 soft drinks are required for every 54 guests. If there are 378 guests, how many soft drinks is required? _____

3) In Mason's class, 42 of the students are tall and 24 are short. In Michael's class 84 students are tall and 48 students are short. Which class has a higher ratio of tall to short students? _____

4) The price of 5 apples at the Quick Market is $4.6. The price of 7 of the same apples at Walmart is $5.95. Which place is the better buy? _____

5) The bakers at a Bakery can make 90 bagels in 3 hours. How many bagels can they bake in 24 hours? What is that rate per hour? _____

6) You can buy 5 cans of green beans at a supermarket for $5.75. How much does it cost to buy 45 cans of green beans? _____

7) The ratio of boys to girls in a class is 4: 7. If there are 32 boys in the class, how many girls are in that class? _____

8) The ratio of red marbles to blue marbles in a bag is 3: 7. If there are 50 marbles in the bag, how many of the marbles are red? _____

AFOQT Subject Test – Mathematics

Percentage Calculations

✎ **Calculate the given percent of each value.**

1) 3% of 60 = ___
2) 20% of 32 = ___
3) 4% of 72 = ___
4) 16% of 32 = ___
5) 25% of 124 = ___
6) 35% of 56 = ___
7) 15% of 20 = ___
8) 14% of 150 = ___
9) 80% of 50 = ___
10) 12% of 115 = ___
11) 72% of 250 = ___
12) 52% of 500 = ___
13) 70% of 400 = ___
14) 27% of 145 = ___
15) 90% of 64 = ___
16) 60% of 55 = ___
17) 22% of 210 = ___
18) 8% of 235 = ___

✎ **Calculate the percent of each given value.**

19) ___% of 25 = 5
20) ___% of 40 = 20
21) ___% of 25 = 2
22) ___% of 50 = 16
23) ___% of 250 = 5
24) ___% of 40 = 32
25) ___% of 125 = 20
26) ___% of 700 = 49
27) ___% of 350 = 49
28) ___% of 500 = 210

✎ **Calculate each percent problem.**

29) A Cinema has 250 seats. 60 seats were sold for the current movie. What percent of seats are empty? ___ %

30) There are 68 boys and 92 girls in a class. 75% of the students in the class take the bus to school. How many students do not take the bus to school? ___

AFOQT Subject Test – Mathematics

Percent Problems

✎ **Calculate each problem.**

1) 9 is what percent of 45? ___%

2) 60 is what percent of 120? ___%

3) 10 is what percent of 200? ___%

4) 15 is what percent of 125? ___%

5) 10 is what percent of 400? ___%

6) 66 is what percent of 55? ___%

7) 40 is what percent of 160? ___%

8) 40 is what percent of 50? ___%

9) 120 is what percent of 800? ___%

10) 78 is what percent of 120? ___%

11) 36 is what percent of 144? ___%

12) 17 is what percent of 85? ___%

13) 90 is what percent of 900? ___%

14) 36 is what percent of 16? ___%

15) 63 is what percent of 14? ___%

16) 18 is what percent of 60? ___%

17) 126 is what percent of 200? ___%

18) 232 is what percent of 40? ___%

✎ **Calculate each percent word problem.**

19) There are 40 employees in a company. On a certain day, 25 were present. What percent showed up for work? ____%

20) A metal bar weighs 60 ounces. 25% of the bar is gold. How many ounces of gold are in the bar? _____

21) A crew is made up of 12 women; the rest are men. If 15% of the crew are women, how many people are in the crew? _____

22) There are 40 students in a class and 8 of them are girls. What percent are boys? ____%

23) The Royals softball team played 400 games and won 280 of them. What percent of the games did they lose? ____%

AFOQT Subject Test – Mathematics

Discount, Tax and Tip

✍ Find the selling price of each item.

1) Original price of a computer: $420
   Tax: 8% Selling price: $_____

2) Original price of a laptop: $280
   Tax: 4% Selling price: $_____

3) Original price of a sofa: $820
   Tax: 5% Selling price: $_____

4) Original price of a car: $15,800
   Tax: 3.6%   Selling price: $_____

5) Original price of a Table: $250
   Tax: 9% Selling price: $_____

6) Original price of a house: $630,000
   Tax: 1.8%   Selling price: $_____

7) Original price of a tablet: $450
   Discount: 30%   Selling price: $____

8) Original price of a chair: $390
   Discount: 8%   Selling price: $____

9) Original price of a book: $75
   Discount: 42%   Selling price: $____

10) Original price of a cellphone: $820
    Discount: 23%   Selling price: $___

11) Food bill: $45
    Tip: 15%   Price: $_____

12) Food bill: $32
    Tipp: 20%   Price: $_____

13) Food bill: $90
    Tip: 35%   Price: $_____

14) Food bill: $42
    Tipp: 12%   Price: $_____

✍ Find the answer for each word problem.

15) Nicolas hired a moving company. The company charged $500 for its services, and Nicolas gives the movers a 40% tip. How much does Nicolas tip the movers? $_____

16) Mason has lunch at a restaurant and the cost of his meal is $90. Mason wants to leave a 25% tip. What is Mason's total bill including tip? $_____

17) The sales tax in Texas is 19.80% and an item costs $350. How much is the tax? $_____

18) The price of a table at Best Buy is $680. If the sales tax is 5%, what is the final price of the table including tax? $_____

WWW.MathNotion.Com

AFOQT Subject Test – Mathematics

Percent of Change

✎ **Find each percent of change.**

1) From 150 to 450. ___ %

2) From 50 ft to 250 ft. ___ %

3) From $60 to $360. ___ %

4) From 60 cm to 180 cm. ___ %

5) From 15 to 45. ___ %

6) From 80 to 16. ___ %

7) From 120 to 360. ___ %

8) From 900 to 450. ___ %

9) From 1,000 to 200. ___ %

10) From 144 to 36. ___ %

✎ **Calculate each percent of change word problem.**

11) Bob got a raise, and his hourly wage increased from $42 to $63. What is the percent increase? ___ %

12) The price of a pair of shoes increases from $50 to $61. What is the percent increase? ___ %

13) At a coffee shop, the price of a cup of coffee increased from $4.80 to $5.76. What is the percent increase in the cost of the coffee? ___ %

14) 51 cm are cut from 85 cm board. What is the percent decrease in length? ___ %

15) In a class, the number of students has been increased from 54 to 81. What is the percent increase? ___ %

16) The price of gasoline rises from $24.40 to $30.50 in one month. By what percent did the gas price rise? ___ %

17) A shirt was originally priced at $38. It went on sale for $24.70. What was the percent that the shirt was discounted? ___ %

AFOQT Subject Test – Mathematics

Simple Interest

✎ **Determine the simple interest for these loans.**

1) $480 at 11% for 3 years. $ _____

2) $4,200 at 7% for 4 years. $ _____

3) $2,500 at 20% for 3 years. $ _____

4) $6,800 at 3.9% for 4 months. $ _____

5) $800 at 6% for 7 months. $ _____

6) $36,000 at 4.2% for 6 years. $ _____

7) $6,500 at 7% for 4 years. $ _____

8) $850 at 9.5% for 2 years. $ _____

9) $1,200 at 5.8% for 9 months. $ ____

10) $3,000 at 4.5% for 7 years. $ ____

✎ **Calculate each simple interest word problem.**

11) A new car, valued at $22,000, depreciates at 8.5% per year. What is the value of the car one year after purchase? $_____

12) Sara puts $9,000 into an investment yielding 6% annual simple interest; she left the money in for three years. How much interest does Sara get at the end of those three years? $_____

13) A bank is offering 12% simple interest on a savings account. If you deposit $16,400, how much interest will you earn in two years? $_____

14) $720 interest is earned on a principal of $6,000 at a simple interest rate of 4% interest per year. For how many years was the principal invested? _____

15) In how many years will $2,200 yield an interest of $440 at 4% simple interest? _____

16) Jim invested $8,000 in a bond at a yearly rate of 4.5%. He earned $1,440 in interest. How long was the money invested? _____

AFOQT Subject Test – Mathematics

Answers of Worksheets

Simplifying Ratios

1) 3 : 4	14) 3 : 4	26) $\frac{3}{5}$	36) $\frac{4}{9}$
2) 1 : 10	15) 1 : 8	27) $\frac{3}{7}$	37) $\frac{1}{12}$
3) 4 : 7	16) 5 : 7	28) $\frac{1}{3}$	38) $\frac{3}{11}$
4) 1 : 3	17) 7 : 9	29) $\frac{1}{3}$	39) $\frac{1}{15}$
5) 1 : 10	18) 2 : 1	30) $\frac{3}{14}$	40) $\frac{1}{4}$
6) 1 : 8	19) 1 : 3	31) $\frac{1}{2}$	41) $\frac{4}{3}$
7) 2 : 8	20) 7 : 1	32) $\frac{1}{5}$	42) $\frac{1}{5}$
8) 2 : 5	21) 1 : 2	33) $\frac{5}{12}$	43) $\frac{22}{41}$
9) 1 : 6	22) 5 : 4	34) $\frac{2}{9}$	44) $\frac{1}{9}$
10) 7 : 9	23) 1 : 2	35) $\frac{11}{13}$	45) $\frac{3}{5}$
11) 2 : 3	24) 1 : 30		
12) 7 : 2	25) $\frac{1}{2}$		
13) 10 : 1			

Proportional Ratios

1) 18	10) 72	19) Yes	28) 42
2) 50	11) 200	20) Yes	29) 63
3) 9	12) 77	21) No	30) 65
4) 14	13) Yes	22) No	31) 20
5) 14	14) Yes	23) Yes	32) 104
6) 56	15) Yes	24) Yes	33) 55
7) 100	16) No	25) 40	34) 64
8) 25	17) No	26) 256	35) 126
9) 70	18) No	27) 7	36) 126

Similarity and ratios

1) 15	4) 13	7) 50 feet
2) 5	5) 150 feet	8) 307.2 miles
3) 15	6) 12.5 meters	

Ratio and Rates Word Problems

1) 2 : 3   2) 315

AFOQT Subject Test – Mathematics

3) The ratio for both classes is 7 to 4.
4) Walmart is a better buy.
5) 720, the rate is 30 per hour.

6) $51.75
7) 56
8) 15

Percentage Calculations

1) 1.8
2) 6.4
3) 2.88
4) 5.12
5) 31
6) 19.6
7) 3
8) 21
9) 40
10) 13.8

11) 180
12) 260
13) 280
14) 39.15
15) 57.6
16) 33
17) 46.2
18) 18.8
19) 20%
20) 50%

21) 8%
22) 32%
23) 2%
24) 80%
25) 16%
26) 7%
27) 14%
28) 42%
29) 76%
30) 40

Percent Problems

1) 20%
2) 50%
3) 5%
4) 12%
5) 2.5%
6) 120%
7) 25%
8) 80%

9) 15%
10) 65%
11) 25%
12) 20%
13) 10%
14) 225%
15) 450%
16) 30%

17) 63%
18) 580%
19) 62.5%
20) 15 ounces
21) 80
22) 80%
23) 30%

Discount, Tax and Tip

1) $453.60
2) $291.20
3) $861.00
4) $16,368.80
5) $272.50
6) $641,340

7) $315.00
8) $358.80
9) $43.50
10) $631.40
11) $51.75
12) $38.40

13) $121.50
14) $47.04
15) $200.00
16) $112.50
17) $69.30
18) $714.00

AFOQT Subject Test – Mathematics

Percent of Change

1) 200%
2) 400%
3) 500%
4) 200%
5) 200%
6) 80%
7) 200%
8) 50%
9) 80%
10) 75%
11) 50%
12) 22%
13) 20%
14) 60%
15) 50%
16) 25%
17) 35%

Simple Interest

1) $158.40
2) $1,176.00
3) $1,500.00
4) $88.40
5) $28.00
6) $9,072.00
7) $1,820.00
8) $161.50
9) $52.20
10) $945.00
11) $20,130.00
12) $1,620.00
13) $3,936.00
14) 3 years
15) 5 years
16) 4 years

AFOQT Subject Test – Mathematics

Chapter 4: Exponents and Radicals Expressions

Topics that you'll practice in this chapter:

- ✓ Multiplication Property of Exponents
- ✓ Zero and Negative Exponents
- ✓ Division Property of Exponents
- ✓ Powers of Products and Quotients
- ✓ Negative Exponents and Negative Bases
- ✓ Scientific Notation
- ✓ Square Roots
- ✓ Simplifying Radical Expressions

Mathematics is no more computation than typing is literature.
– John Allen Paulos

AFOQT Subject Test – Mathematics

Multiplication Property of Exponents

✎ Simplify and write the answer in exponential form.

1) $4 \times 4^5 =$

2) $8^4 \times 8 =$

3) $7^3 \times 7^3 =$

4) $9^2 \times 9^2 =$

5) $2^2 \times 2^4 \times 2 =$

6) $5 \times 5^3 \times 5^3 =$

7) $4^3 \times 4^2 \times 4 \times 4 =$

8) $5x \times x =$

9) $x^3 \times x^3 =$

10) $x^7 \times x^2 =$

11) $x^4 \times x^3 \times x^2 =$

12) $10x \times 3x =$

13) $4x^3 \times 4x^3 =$

14) $7x^3 \times x =$

15) $3x^2 \times 4x^2 \times x^2 =$

16) $5x^4 \times x^4 =$

17) $2x^8 \times 2x =$

18) $6x \times x^5 =$

19) $4x^2 \times 6x^6 =$

20) $5yx^3 \times 4x =$

21) $7x^3 \times y^5 x^7 =$

22) $y^2 x^3 \times y^5 x^4 =$

23) $3x^5 \times 4x^3 y^4 =$

24) $4x^4 \times 9x^2 y^5 =$

25) $5x^3 y^4 \times 6x^8 y^2 =$

26) $8x^3 y^6 \times 4xy^3 =$

27) $2xy^5 \times 6x^3 y^3 =$

28) $4x^5 y^2 \times 4x^2 y^8 =$

29) $7x \times 3y^8 x^2 \times y^5 =$

30) $x^3 \times 2y^3 x^4 \times 2y =$

31) $3yx^4 \times 3y^4 x \times 3xy^3 =$

32) $6y^3 \times 2y^2 x^4 \times 10yx^5 =$

WWW.MathNotion.Com

AFOQT Subject Test – Mathematics

Zero and Negative Exponents

✏️ **Evaluate the following expressions.**

1) $1^{-5} =$

2) $4^{-1} =$

3) $0^{10} =$

4) $1^{15} =$

5) $5^{-2} =$

6) $3^{-3} =$

7) $9^{-1} =$

8) $10^{-2} =$

9) $12^{-2} =$

10) $2^{-5} =$

11) $3^{-4} =$

12) $2^{-4} =$

13) $6^{-3} =$

14) $10^{-3} =$

15) $30^{-1} =$

16) $15^{-2} =$

17) $4^{-3} =$

18) $2^{-7} =$

19) $5^{-3} =$

20) $4^{-4} =$

21) $3^{-5} =$

22) $10^{-4} =$

23) $2^{-10} =$

24) $8^{-3} =$

25) $20^{-2} =$

26) $14^{-2} =$

27) $9^{-3} =$

28) $100^{-2} =$

29) $5^{-4} =$

30) $4^{-6} =$

31) $(\frac{1}{4})^{-3} =$

32) $(\frac{1}{6})^{-2} =$

33) $(\frac{1}{7})^{-2} =$

34) $(\frac{2}{3})^{-3} =$

35) $(\frac{1}{13})^{-2} =$

36) $(\frac{7}{12})^{-2} =$

37) $(\frac{1}{6})^{-3} =$

38) $(\frac{1}{300})^{-2} =$

39) $(\frac{2}{9})^{-2} =$

40) $(\frac{7}{5})^{-1} =$

41) $(\frac{13}{23})^{0} =$

42) $(\frac{1}{4})^{-5} =$

WWW.MathNotion.Com

AFOQT Subject Test – Mathematics

Division Property of Exponents

✏️ **Simplify.**

1) $\dfrac{5^6}{5^7} =$

2) $\dfrac{8^8}{8^6} =$

3) $\dfrac{4^5}{4} =$

4) $\dfrac{3}{3^5} =$

5) $\dfrac{x}{x^6} =$

6) $\dfrac{3 \times 3^2}{3^2 \times 3^5} =$

7) $\dfrac{9^4}{9^2} =$

8) $\dfrac{10 \times 10^9}{10^2 \times 10^7} =$

9) $\dfrac{7^5 \times 7^7}{7^4 \times 7^8} =$

10) $\dfrac{15x}{30x^6} =$

11) $\dfrac{3x^9}{4x^4} =$

12) $\dfrac{15x^8}{10x^9} =$

13) $\dfrac{42x^5}{6y^9} =$

14) $\dfrac{36y^8}{4x^4y^5} =$

15) $\dfrac{2x^7}{9x} =$

16) $\dfrac{49x^8y^6}{7x^9} =$

17) $\dfrac{48x^2}{24x^6y^{12}} =$

18) $\dfrac{30yx^5}{6yx^7} =$

19) $\dfrac{19x^7y}{38x^{12}y^4} =$

20) $\dfrac{9x^8}{63x^8} =$

21) $\dfrac{9x^{-9}}{4x^{-3}} =$

WWW.MathNotion.Com

AFOQT Subject Test – Mathematics

Powers of Products and Quotients

✍ **Simplify.**

1) $(4^3)^2 =$

2) $(2^3)^4 =$

3) $(2 \times 2^3)^2 =$

4) $(5 \times 5^5)^6 =$

5) $(19^4 \times 19^2)^3 =$

6) $(2^3 \times 2^4)^4 =$

7) $(5 \times 5^2)^2 =$

8) $(4^4)^4 =$

9) $(8x^5)^2 =$

10) $(3x^2y^4)^4 =$

11) $(7x^5y^2)^2 =$

12) $(5x^4y^4)^3 =$

13) $(2x^3y^3)^5 =$

14) $(10x^3y^4)^3 =$

15) $(13y^3y)^2 =$

16) $(5x^6x^4)^2 =$

17) $(6x^7y^6)^3 =$

18) $(12x^5x^7)^2 =$

19) $(2x^4 \times 2x)^4 =$

20) $(2x^4y^3)^5 =$

21) $(15x^7y^2)^2 =$

22) $(8x^3y^5)^3 =$

23) $(3x \times 2y^2)^4 =$

24) $\left(\dfrac{4x}{x^5}\right)^2 =$

25) $\left(\dfrac{x^4y^5}{x^3y^5}\right)^9 =$

26) $\left(\dfrac{36xy}{6x^5}\right)^3 =$

27) $\left(\dfrac{x^7}{x^8y^2}\right)^6 =$

28) $\left(\dfrac{xy^4}{x^3y^6}\right)^{-3} =$

29) $\left(\dfrac{5xy^8}{x^3}\right)^2 =$

30) $\left(\dfrac{xy^6}{2xy^3}\right)^{-4} =$

WWW.MathNotion.Com

AFOQT Subject Test – Mathematics

Negative Exponents and Negative Bases

✎ **Simplify.**

1) $-9^{-1} =$

2) $-9^{-2} =$

3) $-2^{-5} =$

4) $-x^{-7} =$

5) $11x^{-1} =$

6) $-8x^{-3} =$

7) $-12x^{-5} =$

8) $-9x^{-8}y^{-6} =$

9) $32x^{-5}y^{-1} =$

10) $10a^{-9}b^{-3} =$

11) $-17x^4y^{-6} =$

12) $-\dfrac{25}{x^{-5}} =$

13) $-\dfrac{13x}{a^{-7}} =$

14) $\left(-\dfrac{1}{3}\right)^{-4} =$

15) $\left(-\dfrac{3}{4}\right)^{-2} =$

16) $-\dfrac{14}{a^{-6}b^{-3}} =$

17) $-\dfrac{7x}{x^{-8}} =$

18) $-\dfrac{a^{-9}}{b^{-5}} =$

19) $-\dfrac{11}{x^{-5}} =$

20) $\dfrac{8b}{-16c^{-6}} =$

21) $\dfrac{12ab}{a^{-4}b^{-3}} =$

22) $-\dfrac{8n^{-4}}{32p^{-7}} =$

23) $\dfrac{16ab^{-6}}{-6c^{-5}} =$

24) $\left(\dfrac{10a}{5c}\right)^{-4} =$

25) $\left(-\dfrac{12x}{4yz}\right)^{-3} =$

26) $\dfrac{8ab^{-7}}{-5c^{-3}} =$

27) $\left(-\dfrac{x^4}{x^5}\right)^{-5} =$

28) $\left(-\dfrac{x^{-2}}{7x^3}\right)^{-2} =$

29) $\left(-\dfrac{x^{-4}}{x^2}\right)^{-6} =$

WWW.MathNotion.Com

AFOQT Subject Test – Mathematics

Scientific Notation

✎ Write each number in scientific notation.

1) $0.223 =$

2) $0.09 =$

3) $4.5 =$

4) $900 =$

5) $2,000 =$

6) $0.006 =$

7) $33 =$

8) $9,400 =$

9) $1,470 =$

10) $52,000 =$

11) $8,000,000 =$

12) $0.00009 =$

13) $2,158,000 =$

14) $0.0039 =$

15) $0.000075 =$

16) $4,300,000 =$

17) $130,000 =$

18) $4,000,000,000 =$

19) $0.00009 =$

20) $0.0039 =$

✎ Write each number in standard notation.

21) $4 \times 10^{-1} =$

22) $1.2 \times 10^{-3} =$

23) $2.7 \times 10^5 =$

24) $6 \times 10^{-4} =$

25) $3.6 \times 10^{-3} =$

26) $5.5 \times 10^5 =$

27) $3.2 \times 10^4 =$

28) $3.88 \times 10^6 =$

29) $7 \times 10^{-6} =$

30) $4.2 \times 10^{-7} =$

AFOQT Subject Test – Mathematics

Square Roots

✏ Find the value each square root.

1) $\sqrt{16}=$ ___

2) $\sqrt{25}=$ ___

3) $\sqrt{1}=$ ___

4) $\sqrt{64}=$ ___

5) $\sqrt{0}=$ ___

6) $\sqrt{196}=$ ___

7) $\sqrt{4}=$ ___

8) $\sqrt{256}=$ ___

9) $\sqrt{36}=$ ___

10) $\sqrt{289}=$ ___

11) $\sqrt{169}=$ ___

12) $\sqrt{144}=$ ___

13) $\sqrt{100}=$ ___

14) $\sqrt{1,600}=$ ___

15) $\sqrt{2,500}=$ ___

16) $\sqrt{324}=$ ___

17) $\sqrt{529}=$ ___

18) $\sqrt{20}=$ ___

19) $\sqrt{625}=$ ___

20) $\sqrt{18}=$ ___

21) $\sqrt{50}=$ ___

22) $\sqrt{1,024}=$ ___

23) $\sqrt{160}=$ ___

24) $\sqrt{32}=$ ___

✏ Evaluate.

25) $\sqrt{4}\times\sqrt{25}=$ _____

26) $\sqrt{36}\times\sqrt{49}=$ _____

27) $\sqrt{6}\times\sqrt{6}=$ _____

28) $\sqrt{13}\times\sqrt{13}=$ _____

29) $2\sqrt{5}\times3\sqrt{5}=$ _____

30) $\sqrt{12}\times\sqrt{3}=$ _____

31) $\sqrt{13}+\sqrt{13}=$ _____

32) $\sqrt{10}+2\sqrt{10}=$ _____

33) $12\sqrt{7}-10\sqrt{7}=$ _____

34) $4\sqrt{10}\times2\sqrt{10}=$ _____

35) $5\sqrt{3}\times8\sqrt{3}=$ _____

36) $6\sqrt{3}-\sqrt{12}=$ _____

WWW.MathNotion.Com

AFOQT Subject Test – Mathematics

Simplifying Radical Expressions

✎ **Simplify.**

1) $\sqrt{13x^2} =$

2) $\sqrt{75x^2} =$

3) $\sqrt[3]{27a} =$

4) $\sqrt{64x^5} =$

5) $\sqrt{216a} =$

6) $\sqrt[3]{63w^3} =$

7) $\sqrt{192x} =$

8) $\sqrt{125v} =$

9) $\sqrt[3]{128x^2} =$

10) $\sqrt{100x^9} =$

11) $\sqrt{16x^4} =$

12) $\sqrt[3]{500a^5} =$

13) $\sqrt{242} =$

14) $\sqrt{392p^3} =$

15) $\sqrt{8m^6} =$

16) $\sqrt{198x^3y^3} =$

17) $\sqrt{121x^5y^5} =$

18) $\sqrt{16a^6b^3} =$

19) $\sqrt{90x^5y^7} =$

20) $\sqrt[3]{64y^2x^6} =$

21) $10\sqrt{16x^4} =$

22) $6\sqrt{81x^2} =$

23) $\sqrt[3]{56x^2y^6} =$

24) $\sqrt[3]{1{,}000x^5y^7} =$

25) $8\sqrt{50a} =$

26) $\sqrt[4]{625x^8y} =$

27) $\sqrt{24x^4y^5r^3} =$

28) $5\sqrt{36x^4y^5z^8} =$

29) $3\sqrt[3]{343x^9y^7} =$

30) $5\sqrt{81a^5b^2c^9} =$

31) $\sqrt[4]{625x^8y^{16}} =$

WWW.MathNotion.Com

AFOQT Subject Test – Mathematics

Answers of Worksheets

Multiplication Property of Exponents

1) 4^6
2) 8^5
3) 7^6
4) 9^4
5) 2^7
6) 5^7
7) 4^7
8) $5x^2$
9) x^6
10) x^9
11) x^9
12) $30x^2$
13) $16x^6$
14) $7x^4$
15) $12x^6$
16) $5x^8$
17) $4x^9$
18) $6x^6$
19) $24x^8$
20) $20x^4 y$
21) $7x^{10} y^5$
22) $x^7 y^7$
23) $12x^8 y^4$
24) $36x^6 y^5$
25) $30x^{11} y^6$
26) $32x^4 y^9$
27) $12x^4 y^8$
28) $16x^7 y^{10}$
29) $21x^3 y^{13}$
30) $4x^7 y^4$
31) $27x^6 y^8$
32) $120x^9 y^6$

Zero and Negative Exponents

1) 1
2) $\frac{1}{4}$
3) 0
4) 1
5) $\frac{1}{25}$
6) $\frac{1}{27}$
7) $\frac{1}{9}$
8) $\frac{1}{100}$
9) $\frac{1}{144}$
10) $\frac{1}{32}$
11) $\frac{1}{81}$
12) $\frac{1}{16}$
13) $\frac{1}{216}$
14) $\frac{1}{1,000}$
15) $\frac{1}{30}$
16) $\frac{1}{225}$
17) $\frac{1}{64}$
18) $\frac{1}{128}$
19) $\frac{1}{125}$
20) $\frac{1}{256}$
21) $\frac{1}{243}$
22) $\frac{1}{10,000}$
23) $\frac{1}{1,024}$
24) $\frac{1}{512}$
25) $\frac{1}{400}$
26) $\frac{1}{196}$
27) $\frac{1}{729}$
28) $\frac{1}{10,000}$
29) $\frac{1}{625}$
30) $\frac{1}{4,096}$
31) 64
32) 36
33) 49
34) $\frac{27}{8}$
35) 169
36) $\frac{144}{49}$
37) 216
38) $90,000$
39) $\frac{81}{4}$
40) $\frac{5}{7}$
41) 1
42) $1,024$

Division Property of Exponents

1) $\frac{1}{5}$
2) 8^2
3) 4^4
4) $\frac{1}{3^4}$
5) $\frac{1}{x^5}$
6) $\frac{1}{3^4}$
7) 9^2
8) 10
9) 1
10) $\frac{1}{2x^5}$
11) $\frac{3x^5}{4}$
12) $\frac{3}{2x}$
13) $\frac{7x^5}{y^9}$
14) $\frac{9y^3}{x^4}$

AFOQT Subject Test – Mathematics

15) $\frac{2x^6}{9}$
16) $\frac{7y^6}{x}$
17) $\frac{2}{x^4 y^{12}}$
18) $\frac{5}{x^2}$
19) $\frac{1}{2x^5 y^3}$
20) $\frac{1}{7}$
21) $\frac{9}{4x^6}$

Powers of Products and Quotients

1) 4^6
2) 2^{12}
3) 2^8
4) 5^{36}
5) 19^{18}
6) 2^{28}
7) 5^6
8) 4^{16}
9) $64x^{10}$
10) $81x^8 y^{16}$
11) $49x^{10} y^4$
12) $125x^{12} y^{12}$
13) $32x^{15} y^{15}$
14) $1,000x^9 y^{12}$
15) $169y^8$
16) $25x^{20}$
17) $216x^{21} y^{18}$
18) $144x^{24}$
19) $256x^{20}$
20) $32x^{20} y^{15}$
21) $225x^{14} y^4$
22) $512x^9 y^{15}$
23) $1,296x^4 y^8$
24) $\frac{16}{x^8}$
25) x^9
26) $\frac{216y^3}{x^{12}}$
27) $\frac{1}{x^6 y^{12}}$
28) $x^6 y^6$
29) $\frac{25y^{16}}{x^4}$
30) $\frac{16}{y^{12}}$

Negative Exponents and Negative Bases

1) $-\frac{1}{9}$
2) $-\frac{1}{81}$
3) $-\frac{1}{32}$
4) $-\frac{1}{x^7}$
5) $\frac{11}{x}$
6) $-\frac{8}{x^3}$
7) $-\frac{12}{x^5}$
8) $-\frac{9}{x^8 y^6}$
9) $\frac{32}{x^5 y}$
10) $\frac{10}{a^9 b^3}$
11) $-\frac{17x^4}{y^6}$
12) $-25x^5$
13) $-13xa^7$
14) 81
15) $\frac{16}{9}$
16) $-14a^6 b^3$
17) $-7x^9$
18) $-\frac{b^5}{a^9}$
19) $-11x^5$
20) $-\frac{bc^6}{2}$
21) $12a^5 b^4$
22) $-\frac{p^7}{4n^4}$
23) $-\frac{8ac^5}{3b^6}$
24) $\frac{c^4}{16a^4}$
25) $\frac{y^3 z^3}{27x^3}$
26) $-\frac{8ac^3}{5b^7}$
27) $-x^5$
28) $49x^{10}$
29) x^{36}

Scientific Notation

1) 2.23×10^{-1}
2) 9×10^{-2}
3) 4.5×10^0
4) 9×10^2
5) 2×10^3
6) 6×10^{-3}
7) 3.3×10^1
8) 9.4×10^3
9) 1.47×10^3

WWW.MathNotion.Com

AFOQT Subject Test – Mathematics

10) 5.2×10^4
11) 8×10^6
12) 9×10^{-5}
13) 2.158×10^6
14) 3.9×10^{-3}
15) 7.5×10^{-5}
16) 4.3×10^6

17) 1.3×10^5
18) 4×10^9
19) 9×10^{-5}
20) 3.9×10^{-3}
21) 0.4
22) 0.0012
23) $270,000$

24) 0.0006
25) 0.0036
26) $550,000$
27) $32,000$
28) $3,880,000$
29) 0.000007
30) 0.00000042

Square Roots

1) 4
2) 5
3) 1
4) 8
5) 0
6) 14
7) 2
8) 16
9) 6

10) 17
11) 13
12) 12
13) 10
14) 40
15) 50
16) 18
17) 23
18) $2\sqrt{5}$

19) 25
20) $3\sqrt{2}$
21) $5\sqrt{2}$
22) 32
23) $4\sqrt{10}$
24) $4\sqrt{2}$
25) 10
26) 42
27) 6

28) 13
29) 30
30) 6
31) $2\sqrt{13}$
32) $3\sqrt{10}$
33) $2\sqrt{7}$
34) 80
35) 120
36) $4\sqrt{3}$

Simplifying radical expressions

1) $x\sqrt{13}$
2) $5x\sqrt{3}$
3) $3\sqrt[3]{a}$
4) $8x^2\sqrt{x}$
5) $6\sqrt{6a}$
6) $w\sqrt[3]{63}$
7) $8\sqrt{3x}$
8) $5\sqrt{5v}$
9) $4\sqrt[3]{2x^2}$
10) $10x^4\sqrt{x}$
11) $4x^2$

12) $5a\sqrt[3]{4a^2}$
13) $11\sqrt{2}$
14) $14p\sqrt{2p}$
15) $2m^3\sqrt{2}$
16) $3x.y\sqrt{22xy}$
17) $11x^2y^2\sqrt{xy}$
18) $4a^3b\sqrt{b}$
19) $3x^2y^3\sqrt{10xy}$
20) $4x^2\sqrt[3]{y^2}$
21) $40x^2$
22) $54x$

23) $2y^2\sqrt[3]{7x^2}$
24) $10xy^2\sqrt[3]{x^2y}$
25) $40\sqrt{2a}$
26) $5x^2\sqrt[4]{y}$
27) $2x^2y^2r\sqrt{6yr}$
28) $30x^2y^2z^4\sqrt{y}$
29) $21x^3y^2\sqrt[3]{y}$
30) $45a^2bc^4\sqrt{ac}$
31) $5x^2y^4$

AFOQT Subject Test – Mathematics

Chapter 5 :
Algebraic Expressions

Topics that you'll practice in this chapter:

- ✓ Simplifying Variable Expressions
- ✓ Simplifying Polynomial Expressions
- ✓ Translate Phrases into an Algebraic Statement
- ✓ The Distributive Property
- ✓ Evaluating One Variable Expressions
- ✓ Evaluating Two Variables Expressions
- ✓ Combining like Terms

Mathematics is, as it were, a sensuous logic, and relates to philosophy as do the arts, music, and plastic art to poetry. — *K. Shegel*

AFOQT Subject Test – Mathematics

Simplifying Variable Expressions

✎ **Simplify each expression.**

1) $3(x + 5) =$

2) $(-4)(7x - 5) =$

3) $11x + 5 - 6x =$

4) $-4 - 2x^2 - 6x^2 =$

5) $7 + 13x^2 + 3 =$

6) $3x^2 + 7x + 15x^2 =$

7) $3x^2 - 12x^2 + 4x =$

8) $4x^2 - 8x - 2x =$

9) $6x + 7(3 - 4x) =$

10) $8x + 4(15x - 3) =$

11) $6(-3x - 9) - 17 =$

12) $-11x^2 - (-5x) =$

13) $2x + 7 + 5 - 8x =$

14) $7 + 6x - 11 - 5x =$

15) $27x + 8 - 13 - 5x =$

16) $(-11)(-5x + 2) - 41x =$

17) $19x - 4(4 - 2x) =$

18) $16x + 3(3x + 6) + 10 =$

19) $5(-2x - 4) - 13x =$

20) $16x - 3x(x + 10) =$

21) $17x + 5x(2 - 4x) =$

22) $5x(-4x - 7) + 20x =$

23) $25x - 19 + 4x^2 =$

24) $6x(x - 11) + 25 =$

25) $4x - 5 + 15x + 3x^2 =$

26) $-7x^2 - 11x - 9x =$

27) $10x - 9x^2 - 3x^2 - 7 =$

28) $13 + 3x^2 - 9x^2 - 21x =$

29) $22x + 10x^2 - 15x + 17 =$

30) $4x^2 + 25x + 21x^2 =$

31) $29 - 12x^2 - 23x - 4x^2 =$

32) $22x - 19x - 9x^2 + 30 =$

Simplifying Polynomial Expressions

✏ **Simplify each polynomial.**

1) $(2x^3 + 8x^2) - (11x + 3x^2) =$ _____

2) $(2x^5 + 7x^3) - (5x^3 + 11x^2) =$ _____

3) $(41x^4 + 5x^2) - (4x^2 + 20x^4) =$ _____

4) $13x - 8x^2 + 4(4x^2 + 3x^3) =$ _____

5) $(4x^3 - 22) + 5(3x^2 - 6x^3) =$ _____

6) $(4x^3 - 3x) - 5(2x^3 + x^4) =$ _____

7) $5(5x - 2x^3) - 2(8x^3 + 5x^2) =$ _____

8) $(3x^2 - 10x) - (5x^3 + 14x^2) =$ _____

9) $5x^3 - (3x^4 + 5x) + 2x^2 =$ _____

10) $11x^4 - (3x^2 + 5x) + 7x =$ _____

11) $(6x^2 - 3x^4) - (10x^4 + 3x^2) =$ _____

12) $2x^2 - 7x^3 + 19x^4 - 22x^3 =$ _____

13) $10x^2 - x^4 + 4x^4 - 32x^3 =$ _____

14) $-5x^2 + 17x^3 - 8x^2 - 6x =$ _____

15) $x^4 - 11x^5 - 30x^4 + 5x^2 =$ _____

16) $21x^3 + 13x - 5x^2 - 11x^3 =$ _____

AFOQT Subject Test – Mathematics

Translate Phrases into an Algebraic Statement

✎ Write an algebraic expression for each phrase.

1) 9 multiplied by x. _____

2) Subtract 11 from y. _____

3) 19 divided by x. _____

4) 38 decreased by y. _____

5) Add y to 40. _____

6) The square of 6. _____

7) x raised to the fifth power. _____

8) The sum of six and a number. _____

9) The difference between fifty–seven and y. _____

10) The quotient of nine and a number. _____

11) The quotient of the square of x and 25. _____

12) The difference between x and 6 is 19. _____

13) 10 times a reduced by the square of b. _____

14) Subtract the product of a and b from 41. _____

AFOQT Subject Test – Mathematics

The Distributive Property

✎ Use the distributive property to simply each expression.

1) $4(1 + 2x) =$

2) $2(4 + 7x) =$

3) $3(4x - 4) =$

4) $(2x - 5)(-6) =$

5) $(-3)(x + 6) =$

6) $(4 + 3x)2 =$

7) $(-5)(8 - 3x) =$

8) $-(-5 - 7x) =$

9) $(-6x + 3)(-3) =$

10) $(-4)(x - 7) =$

11) $-(5 - 3x) =$

12) $3(9 + 4x) =$

13) $6(4 + 3x) =$

14) $(-5x + 3)2 =$

15) $(5 - 8x)(-3) =$

16) $(-12)(3x + 3) =$

17) $(5 - 3x)6 =$

18) $4(2 + 6x) =$

19) $8(7x - 3) =$

20) $(-2x + 3)4 =$

21) $(7 - 5x)(-9) =$

22) $(-10)(x - 8) =$

23) $(11 - 4x)3 =$

24) $(-6)(10x - 4) =$

25) $(3 - 9x)(-7) =$

26) $(-9)(x + 9) =$

27) $(-3 + 5x)(-7) =$

28) $(-5)(8 - 10x) =$

29) $12(4x - 8) =$

30) $(-10x + 13)(-3) =$

31) $(-8)(3x - 2) + 4(x + 5) =$

32) $(-8)(x + 4) - (6 + 5x) =$

WWW.MathNotion.Com

AFOQT Subject Test – Mathematics

Evaluating One Variable Expressions

✏️ **Evaluate each expression using the value given.**

1) $8 - x$, $x = 5$

2) $x - 9$, $x = 5$

3) $5x + 4$, $x = 3$

4) $x - 13$, $x = -4$

5) $12 - x$, $x = 4$

6) $x + 2$, $x = 6$

7) $4x + 8$, $x = 3$

8) $x + (-7)$, $x = -8$

9) $4x + 5$, $x = 2$

10) $3x + 9$, $x = -2$

11) $15 + 3x - 7$, $x = 2$

12) $17 - 3x$, $x = 3$

13) $8x - 9$, $x = 4$

14) $5x + 4$, $x = -3$

15) $10x + 5$, $x = 3$

16) $14 - 4x$, $x = -6$

17) $3(5x + 3)$, $x = 9$

18) $4(-3x - 6)$, $x = 3$

19) $7x - 2x + 12$, $x = 4$

20) $(5x + 6) \div 2$, $x = 8$

21) $(x + 18) \div 10$, $x = 12$

22) $5x - 12 + 3x$, $x = -3$

23) $(6 - 4x)(-3)$, $x = -4$

24) $9x^2 + 3x - 6$, $x = 2$

25) $x^2 - 10x$, $x = -5$

26) $3x(7 - 2x)$, $x = 2$

27) $12x + 6 - 2x^2$, $x = -4$

28) $(-3)(4x - 8 + 3x)$, $x = 3$

29) $(-6) + \frac{x}{4} + 3x$, $x = 16$

30) $(-6) + \frac{x}{5}$, $x = 35$

31) $\left(-\frac{45}{x}\right) - 7 + 2x$, $x = 9$

32) $\left(-\frac{21}{x}\right) - 12 + 4x$, $x = 7$

AFOQT Subject Test – Mathematics

Evaluating Two Variables Expressions

✎ **Evaluate each expression using the values given.**

1) $2x - 4y$,
   $x = 4, y = 1$

2) $3x + 5y$,
   $x = -2, y = 2$

3) $-7a + 4b$,
   $a = 2, b = 4$

4) $3x + 5 - y$,
   $x = 5, y = 6$

5) $3z + 12 - 2k$,
   $z = 5, k = 6$

6) $6(-x - 3y)$,
   $x = 5, y = -2$

7) $5a + 3b$,
   $a = 3, b = 4$

8) $7x \div 3y$,
   $x = 3, y = 7$

9) $2x + 15 + 5y$,
   $x = -3, y = 1$

10) $5a - (18 - b)$,
    $a = 2, b = 8$

11) $2z + 20 + 5k$,
    $z = -6, k = 5$

12) $xy + 10 + 4x$,
    $x = 3, y = 5$

13) $2x + 4y - 8 + 5$,
    $x = 5, y = 2$

14) $\left(-\frac{24}{x}\right) + 3 + 2y$,
    $x = 4, y = 6$

15) $(-3)(-3a - 3b)$,
    $a = 4, b = 5$

16) $12 + 4x - 7 - y$,
    $x = 3, y = 5$

17) $11x + 5 - 8y + 6$,
    $x = 5, y = 2$

18) $10 + 2(-4x - 5y)$,
    $x = 5, y = 4$

19) $5x + 13 + 6y$,
    $x = 5, y = 6$

20) $10a - (7a + 3b) - 11$,
    $a = 3, b = 8$

WWW.MathNotion.Com

AFOQT Subject Test – Mathematics

Combining like Terms

✎ **Simplify each expression.**

1) $11x + 3x + 6 =$

2) $8(2x - 6) =$

3) $18x - 7x + 11 =$

4) $(-4)(6x - 7) =$

5) $22x - 10x - 5 =$

6) $32x - 13 + 8x =$

7) $15 - (8x - 11) =$

8) $-24x + 17 - 11x =$

9) $12x - 8 - 6x + 9 =$

10) $21x + 5 - 36 + 12x =$

11) $28x + 3x - 11 =$

12) $(-3x + 4)5 =$

13) $2 + 4x + 9x - 8 =$

14) $6(2x - 5x) - 4 =$

15) $4(5x + 11) + 3x =$

16) $x - 14 - 11x =$

17) $5(10 + 9x) - 8x =$

18) $42x + 17 - 23x =$

19) $(-7x) + 19 + 20x =$

20) $(-7x) - 33 + 29x =$

21) $4(5x + 3) - 19x =$

22) $5(6 - 2x) - 15x =$

23) $-24x + (11 - 18x) =$

24) $(-9) - (6)(7x + 3) =$

25) $(-1)(8x - 10) - 21x =$

26) $-36x + 14 + 27x - 5x =$

27) $3(-13x + 6) - 17x =$

28) $-5x - 42 + 32x =$

29) $37x - 19x + 15 - 9x =$

30) $3(5x + 7x) - 31 =$

31) $14 - 6x - 15 - 9x =$

32) $-2(-5x - 7x) + 27x =$

WWW.MathNotion.Com

AFOQT Subject Test – Mathematics

Answers of Worksheets

Simplifying Variable Expressions

1) $3x + 15$
2) $-28x + 20$
3) $5x + 5$
4) $-8x^2 - 4$
5) $13x^2 + 10$
6) $18x^2 + 7x$
7) $-9x^2 + 4x$
8) $4x^2 - 10x$
9) $-22x + 21$
10) $68x - 12$
11) $-18x - 71$
12) $-11x^2 + 5x$
13) $-6x + 12$
14) $x - 4$
15) $22x - 5$
16) $14x - 22$
17) $27x - 16$
18) $25x + 28$
19) $-23x - 20$
20) $-3x^2 - 14x$
21) $-20x^2 + 27x$
22) $-20x^2 - 15x$
23) $4x^2 + 25x - 19$
24) $6x^2 - 66x + 25$
25) $3x^2 + 19x - 5$
26) $-7x^2 - 20x$
27) $-12x^2 + 10x - 7$
28) $-6x^2 - 21x + 13$
29) $10x^2 + 7x + 17$
30) $25x^2 + 25x$
31) $-16x^2 - 23x + 29$
32) $-9x^2 + 3x + 30$

Simplifying Polynomial Expressions

1) $2x^3 + 5x^2 - 11x$
2) $2x^5 + 2x^3 - 11x^2$
3) $21x^4 + x^2$
4) $12x^3 + 8x^2 + 13x$
5) $-26x^3 + 15x^2 - 22$
6) $-5x^4 - 6x^3 - 3x$
7) $-26x^3 - 10x^2 + 25x$
8) $-5x^3 - 11x^2 - 10x$
9) $-3x^4 + 5x^3 + 2x^2 - 5x$
10) $11x^4 - 3x^2 + 2x$
11) $-13x^4 + 3x^2$
12) $19x^4 - 29x^3 + 2x^2$
13) $3x^4 - 32x^3 + 10x^2$
14) $17x^3 - 13x^2 - 6x$
15) $-11x^5 - 29x^4 + 5x^2$
16) $10x^3 - 5x^2 + 13x$

Translate Phrases into an Algebraic Statement

1) $9x$
2) $y - 11$
3) $\frac{19}{x}$
4) $38 - y$
5) $y + 40$
6) 6^2
7) x^5
8) $6 + x$
9) $57 - y$
10) $\frac{9}{x}$
11) $\frac{x^2}{25}$
12) $x - 6 = 19$
13) $10a - b^2$
14) $41 - ab$

The Distributive Property

1) $8x + 4$
2) $14x + 8$
3) $12x - 12$
4) $-12x + 30$
5) $-3x - 18$
6) $6x + 8$
7) $15x - 40$
8) $7x + 5$
9) $18x - 9$
10) $-4x + 28$
11) $3x - 5$
12) $12x + 27$

AFOQT Subject Test – Mathematics

13) $18x + 24$
14) $-10x + 6$
15) $24x - 15$
16) $-36x - 36$
17) $-18x + 30$

18) $24x + 8$
19) $56x - 24$
20) $-8x + 12$
21) $45x - 63$
22) $-10x + 80$

23) $-12x + 33$
24) $-60x + 24$
25) $63x - 21$
26) $-9x - 81$
27) $-35x + 21$

28) $50x - 40$
29) $48x - 96$
30) $30x - 39$
31) $-20x + 36$
32) $-13x - 38$

Evaluating One Variables

1) 3
2) −4
3) 19
4) −17
5) 8
6) 8
7) 20
8) −15

9) 13
10) 3
11) 14
12) 8
13) 23
14) −11
15) 35
16) 38

17) 144
18) −60
19) 32
20) 23
21) 3
22) −36
23) −66
24) 36

25) 75
26) 18
27) −74
28) −39
29) 46
30) 1
31) 6
32) 13

Evaluating Two Variables

1) 4
2) 4
3) 2
4) 14
5) 15

6) 6
7) 27
8) 1
9) 14
10) 0

11) 33
12) 37
13) 15
14) 9
15) 81

16) 12
17) 50
18) −70
19) 74
20) −26

Combining like Terms

1) $14x + 6$
2) $16x - 48$
3) $11x + 11$
4) $-24x + 28$
5) $12x - 5$
6) $40x - 13$
7) $-8x + 26$
8) $-35x + 17$

9) $6x + 1$
10) $33x - 31$
11) $31x - 11$
12) $-15x + 20$
13) $13x - 6$
14) $-18x - 4$
15) $23x + 44$
16) $-10x - 14$

17) $37x + 50$
18) $19x + 17$
19) $13x + 19$
20) $22x - 33$
21) $x + 12$
22) $-25x + 30$
23) $-42x + 11$
24) $-42x - 27$

25) $-29x + 10$
26) $-14x + 14$
27) $-56x + 18$
28) $27x - 42$
29) $9x + 15$
30) $36x - 31$
31) $-15x - 1$
32) $51x$

AFOQT Subject Test – Mathematics

Chapter 6:
Equations and Inequalities

Topics that you'll practice in this chapter:

- ✓ One-Step Equations
- ✓ Multi-Step Equations
- ✓ Graphing Single-Variable Inequalities
- ✓ One-Step Inequalities
- ✓ Multi-Step Inequalities
- ✓ Systems of Equations
- ✓ Systems of Equations Word Problems

"Life is a math equation. In order to gain the most, you have to know how to convert negatives into positives." – Anonymous

AFOQT Subject Test – Mathematics

One–Step Equations

✎ Find the answer for each equation.

1) $3x = 90, x = $ ___

2) $5x = 35, x = $ ___

3) $6x = 24, x = $ ___

4) $24x = 144, x = $ ___

5) $x + 15 = 20, x = $ ___

6) $x - 7 = 4, x = $ ___

7) $x - 9 = 2, x = $ ___

8) $x + 15 = 23, x = $ ___

9) $x - 4 = 13, x = $ ___

10) $12 = 16 + x, x = $ ___

11) $x - 10 = 2, x = $ ___

12) $5 - x = -11, x = $ ___

13) $28 = -6 + x, x = $ ___

14) $x - 20 = -35, x = $ ___

15) $x + 14 = -4, x = $ ___

16) $14 = 28 - x, x = $ ___

17) $7 + x = -7, x = $ ___

18) $x - 16 = 4, x = $ ___

19) $30 = x - 15, x = $ ___

20) $x - 5 = -18, x = $ ___

21) $x - 10 = 24, x = $ ___

22) $x - 20 = -25, x = $ ___

23) $x - 17 = 30, x = $ ___

24) $-70 = x - 28, x = $ ___

25) $x - 9 = 13, x = $ ___

26) $36 = 4x, x = $ ___

27) $x - 35 = 25, x = $ ___

28) $x - 25 = 10, x = $ ___

29) $70 - x = 16, x = $ ___

30) $x - 10 = 14, x = $ ___

31) $17 - x = -13, x = $ ___

32) $x - 9 = -30, x = $ ___

WWW.MathNotion.Com

AFOQT Subject Test – Mathematics

Multi–Step Equations

✎ Find the answer for each equation.

1) $3x + 3 = 9$

2) $-x + 5 = 12$

3) $4x - 8 = 8$

4) $-(3 - x) = 5$

5) $4x - 8 = 16$

6) $12x - 15 = 9$

7) $2x - 18 = 2$

8) $4x + 8 = 16$

9) $24x + 27 = 75$

10) $-14(3 + x) = 14$

11) $-3(2 + x) = 6$

12) $12 = -(x - 7)$

13) $3(3 - x) = 30$

14) $-15 = -(3x + 6)$

15) $40(3 + x) = 40$

16) $5(x - 10) = 25$

17) $-18 = x + 8x$

18) $3x + 25 = -2x - 10$

19) $7(6 + 3x) = -63$

20) $18 - 3x = -4 - 5x$

21) $4 - 6x = 36 + 2x$

22) $15 + 15x = -5 + 5x$

23) $42 = (-6x) - 7 + 7$

24) $21 = 3x - 21 + 4x$

25) $-18 = -6x - 9 + 3x$

26) $5x - 15 = -29 + 6x$

27) $7x - 18 = 4x + 3$

28) $-7 - 4x = 5(4 - x)$

29) $x - 5 = -5(-3 - x)$

30) $13x - 68 = 15x - 102$

31) $-5x - 3 = -3(9 + 3x)$

32) $-2x - 15 = 6x + 17$

WWW.MathNotion.Com

AFOQT Subject Test – Mathematics

Graphing Single–Variable Inequalities

✎ Draw a graph for each inequality.

1) $x > -1$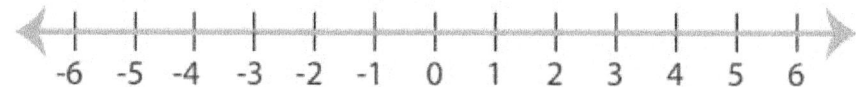

2) $x \leq 2$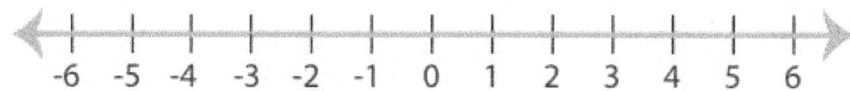

3) $x \geq 0$

4) $x < -3$

5) $x < \frac{1}{2}$

6) $x \leq -2$

7) $x \leq 3$

8) $x \geq -\frac{7}{2}$

WWW.MathNotion.Com

One–Step Inequalities

✏ Find the answer for each inequality and graph it.

1) $x + 4 \geq 4$

2) $x - 5 \leq 2$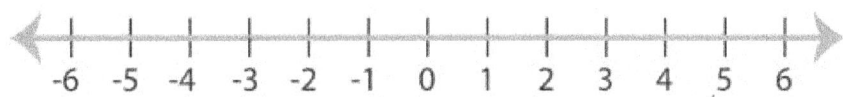

3) $5x > 35$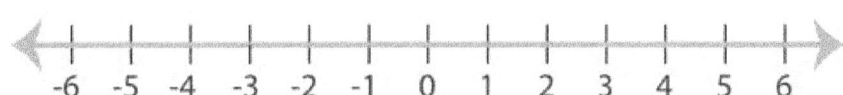

4) $9 + x \leq 11$

5) $x - 5 < -9$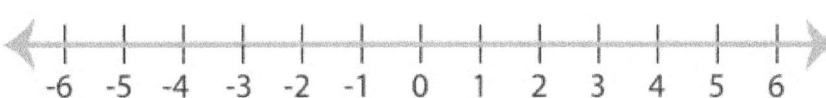

6) $9x \geq 72$

7) $9x \leq 27$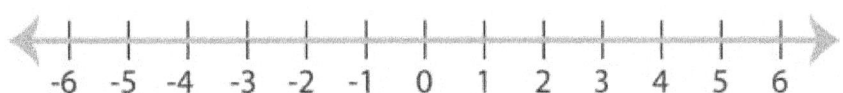

8) $x + 19 > 16$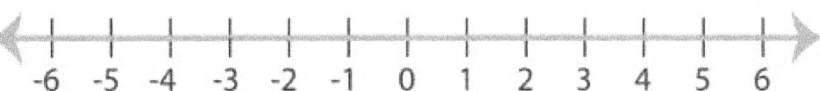

AFOQT Subject Test – Mathematics

Multi-Step Inequalities

✏ **Calculate each inequality.**

1) $x - 3 \leq 7$

2) $8 - x \leq 8$

3) $3x - 9 \leq 9$

4) $4x - 4 \geq 8$

5) $x - 7 \geq 1$

6) $5x - 15 \leq 5$

7) $6x - 8 \leq 4$

8) $-11 + 6x \leq 12$

9) $4(x - 4) \leq 16$

10) $3x - 10 \leq 11$

11) $5x - 25 < 25$

12) $9x - 5 < 22$

13) $20 - 7x \geq -15$

14) $33 + 6x < 45$

15) $8 + 8x \geq 96$

16) $7 + 3x < 13$

17) $4x - 3 < 9$

18) $5(2 - 2x) \geq -30$

19) $-(7 + 6x) < 29$

20) $12 - 8x \geq -20$

21) $-4(x - 6) > 24$

22) $\dfrac{3x + 9}{6} \leq 10$

23) $\dfrac{4x - 10}{3} \leq 2$

24) $\dfrac{2x - 8}{3} > 2$

25) $8 + \dfrac{x}{6} < 9$

26) $\dfrac{9x}{7} - 4 < 5$

27) $\dfrac{15x + 45}{15} > 1$

28) $16 + \dfrac{x}{4} < 6$

Systems of Equations

✏ **Calculate each system of equations.**

1) $-x + y = 2$
   $-4x + 2y = 6$
   $x = \_\_\_$
   $y = \_\_\_$

2) $-15x + 3y = -9$
   $9x - 16y = 48$
   $x = \_\_\_$
   $y = \_\_\_$

3) $y = -7$
   $6x + 5y = 7$
   $x = \_\_\_$
   $y = \_\_\_$

4) $3y = -9x + 15$
   $5x - 4y = -3$
   $x = \_\_\_$
   $y = \_\_\_$

5) $10x - 9y = -13$
   $-5x + 3y = 11$
   $x = \_\_\_$
   $y = \_\_\_$

6) $-12x - 16y = 20$
   $6x - 12y = 30$
   $x = \_\_\_$
   $y = \_\_\_$

7) $5x - 14y = -23$
   $-18x + 21y = 24$
   $x = \_\_\_$
   $y = \_\_\_$

8) $15x - 21y = -6$
   $2x - 3y = -2$
   $x = \_\_\_$
   $y = \_\_\_$

9) $-x + 3y = 3$
   $-14x + 16y = -10$
   $x = \_\_\_$
   $y = \_\_\_$

10) $x + 5y = 50$
    $3x + 10y = 80$
    $x = \_\_\_$
    $y = \_\_\_$

11) $6x - 7y = -8$
    $-x - 4y = -9$
    $x = \_\_\_$
    $y = \_\_\_$

12) $2x + 4y = -10$
    $2x - 8y = 14$
    $x = \_\_\_$
    $y = \_\_\_$

13) $4x + 3y = 12$
    $5x - 3y = 15$
    $x = \_\_\_$
    $y = \_\_\_$

14) $3x - 2y = 3$
    $7x - 8y = 22$
    $x = \_\_\_$
    $y = \_\_\_$

15) $3x + 2y = 5$
    $-10x - 4y = -14$
    $x = \_\_\_$
    $y = \_\_\_$

16) $10x + 7y = 1$
    $-5x - 7y = 24$
    $x = \_\_\_$
    $y = \_\_\_$

AFOQT Subject Test – Mathematics

Systems of Equations Word Problems

✎ **Find the answer for each word problem.**

1) Tickets to a movie cost $4 for adults and $3 for students. A group of friends purchased 8 tickets for $31.00. How many adults ticket did they buy? ____

2) At a store, Eva bought two shirts and five hats for $77.00. Nicole bought three same shirts and four same hats for $84.00. What is the price of each shirt? _____

3) A farmhouse shelters 18 animals, some are pigs, and some are ducks. Altogether there are 66 legs. How many pigs are there? _____

4) A class of 214 students went on a field trip. They took 36 vehicles, some cars and some buses. If each car holds 5 students and each bus hold 22 students, how many buses did they take? _____

5) A theater is selling tickets for a performance. Mr. Smith purchased 5 senior tickets and 3 child tickets for $105 for his friends and family. Mr. Jackson purchased 3 senior tickets and 5 child tickets for $79. What is the price of a senior ticket? $____

6) The difference of two numbers is 10. Their sum is 20. What is the bigger number? $____

7) The sum of the digits of a certain two–digit number is 7. Reversing its digits increase the number by 9. What is the number? _____

8) The difference of two numbers is 11. Their sum is 25. What are the numbers? _____

9) The length of a rectangle is 5 meters greater than 2 times the width. The perimeter of rectangle is 28 meters. What is the length of the rectangle? _____

10) Jim has 25 nickels and dimes totaling $1.80. How many nickels does he have? _____

AFOQT Subject Test – Mathematics

Answers of Worksheets

One–Step Equations

1) 30
2) 7
3) 4
4) 6
5) 5
6) 11
7) 11
8) 8
9) 17
10) −4
11) 12
12) 16
13) 34
14) −15
15) −18
16) 14
17) −14
18) 20
19) 45
20) −13
21) 34
22) −5
23) 47
24) −42
25) 22
26) 9
27) 60
28) 35
29) 54
30) 24
31) 30
32) −21

Multi–Step Equations

1) 2
2) −7
3) 4
4) 8
5) 6
6) 2
7) 10
8) 2
9) 2
10) −4
11) −4
12) −5
13) −7
14) 3
15) −2
16) 15
17) −2
18) −7
19) −5
20) −11
21) −4
22) −2
23) −7
24) 6
25) 3
26) 14
27) 7
28) 27
29) −5
30) 17
31) −6
32) −4

Graphing Single–Variable Inequalities

1)

2)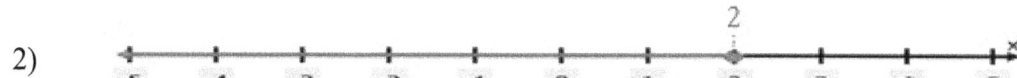

3)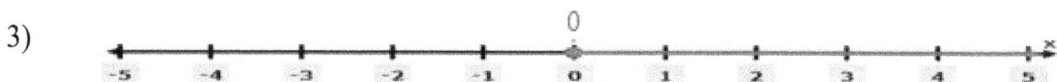

4)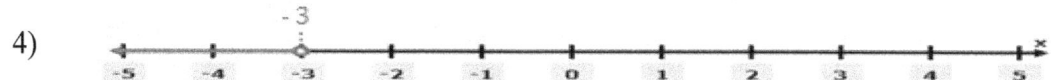

AFOQT Subject Test – Mathematics

5)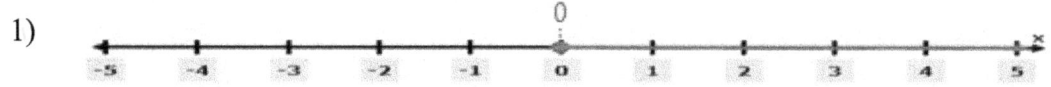

6)

7)

8)

One–Step Inequalities

1)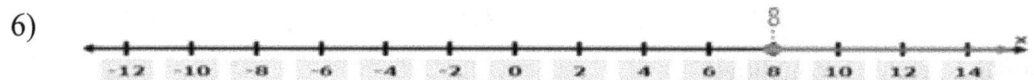

2)

3)

4)

5)

6)

7)

8)

Multi-Step Inequalities

1) $x \leq 10$
2) $x \geq 0$
3) $x \leq 6$
4) $x \geq 3$
5) $x \geq 8$
6) $x \leq 4$
7) $x \leq 2$
8) $x \leq \frac{23}{6}$
9) $x \leq 8$
10) $x \leq 7$
11) $x < 10$
12) $x < 3$
13) $x \leq 5$
14) $x < 2$
15) $x \geq 11$
16) $x < 2$
17) $x < 3$
18) $x \leq 4$
19) $x > -6$
20) $x \leq 4$
21) $x < 0$
22) $x \leq 17$
23) $x \leq 4$

AFOQT Subject Test – Mathematics

24) $x > 7$ 26) $x < 7$ 28) $x < -40$
25) $x < 6$ 27) $x > -2$

Systems of Equations

1) $x = -1, y = 1$
2) $x = 0, y = -3$
3) $x = 7$
4) $x = 1, y = 2$
5) $x = -4, y = -3$
6) $x = 1, y = -2$
7) $x = 1, y = 2$
8) $x = 8, y = 6$
9) $x = 3, y = 2$
10) $x = -20, y = 14$
11) $x = 1, y = 2$
12) $x = -1, y = -2$
13) $x = 3, y = 0$
14) $x = -2, y = -\frac{9}{2}$
15) $x = 1, y = 1$
16) $x = 5, y = -7$

Systems of Equations Word Problems

1) 7
2) $16
3) 15
4) 2
5) $18
6) 15
7) 34
8) 18, 7
9) 11 meters
10) 14

AFOQT Subject Test – Mathematics

Chapter 7:
Linear Functions

Topics that you'll practice in this chapter:

- ✓ Finding Slope
- ✓ Graphing Lines Using Line Equation
- ✓ Writing Linear Equations
- ✓ Graphing Linear Inequalities
- ✓ Finding Midpoint
- ✓ Finding Distance of Two Points

"Nature is written in mathematical language." – Galileo Galilei

AFOQT Subject Test – Mathematics

Finding Slope

✎ Find the slope of each line.

1) $y = x + 8$

2) $y = -3x + 5$

3) $y = 2x + 12$

4) $y = -4x + 19$

5) $y = 11 + 6x$

6) $y = 7 - 5x$

7) $y = 8x + 19$

8) $y = -9x + 20$

9) $y = -7x + 4$

10) $y = 3x - 8$

11) $y = \frac{1}{3}x + 8$

12) $y = -\frac{4}{5}x + 9$

13) $-3x + 6y = 30$

14) $4x + 4y = 16$

15) $3y - x = 10$

16) $8y - x = 5$

✎ Find the slope of the line through each pair of points.

17) $(2, 3), (7, 10)$

18) $(-3, 5), (2, 15)$

19) $(5, -3), (1, 9)$

20) $(-5, -5), (10, 25)$

21) $(22, 3), (7, 18)$

22) $(-16, 8), (-7, 26)$

23) $(25, 11), (29, 19)$

24) $(26, -19), (14, 17)$

25) $(22, -13), (20, -11)$

26) $(19, 7), (15, -3)$

27) $(5, 7), (11, 19)$

28) $(52, -62), (40, 70)$

WWW.MathNotion.Com

AFOQT Subject Test – Mathematics

Graphing Lines Using Line Equation

✎ Sketch the graph of each line.

1) $y = x - 2$

2) $y = -3x + 2$

3) $x + y = 0$

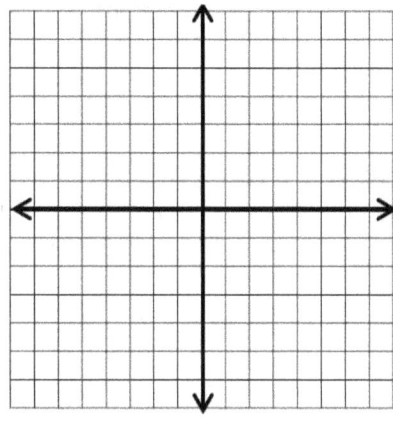
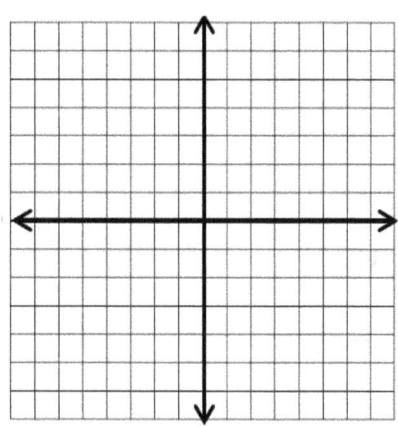
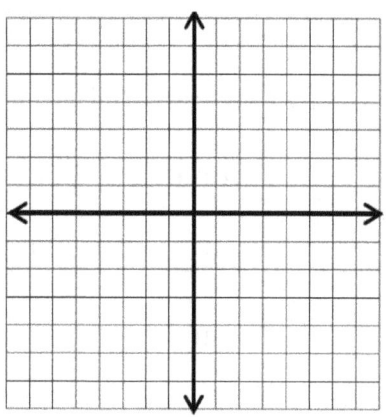

4) $x + y = -3$

5) $2x + 3y = -4$

6) $y - 3x + 6 = 0$

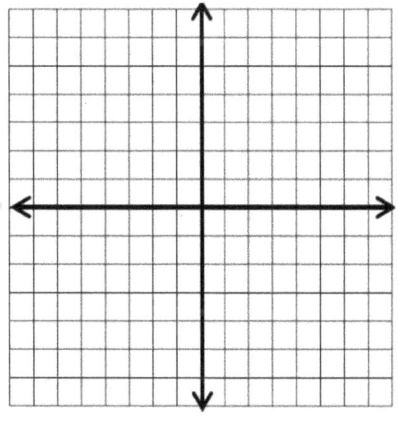
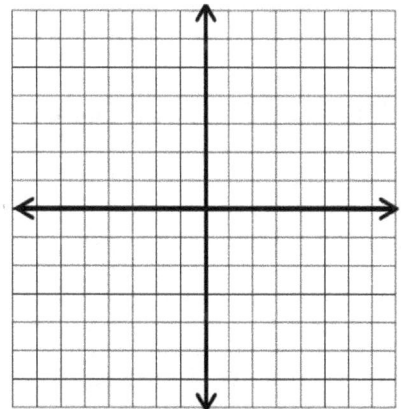
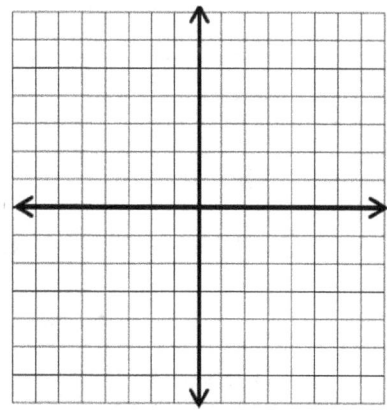

AFOQT Subject Test – Mathematics

Writing Linear Equations

✍ **Write the equation of the line through the given points.**

1) Through: $(2, -5), (3, 9)$

2) Through: $(-6, 3), (3, 12)$

3) Through: $(10, 7), (5, 27)$

4) Through: $(15, 11), (3, -1)$

5) Through: $(24, 17), (12, -7)$

6) Through: $(8, 29), (4, -7)$

7) Through: $(20, -16), (12, 0)$

8) Through: $(-3, 10), (2, -5)$

9) Through: $(-6, 17), (4, -3)$

10) Through: $(-8, 22), (5, -4)$

11) Through: $(9, 27), (3, -3)$

12) Through: $(11, 32), (9, 4)$

13) Through: $(-3, 13), (-4, 0)$

14) Through: $(-5, 5), (5, 15)$

15) Through: $(18, -32), (11, 3)$

16) Through: $(-4, 25), (4, -15)$

✍ **Find the answer for each problem.**

17) What is the equation of a line with slope 6 and intercept 12? _____

18) What is the equation of a line with slope -11 and intercept -4? _____

19) What is the equation of a line with slope -3 and passes through point $(5, 2)$? _____

20) What is the equation of a line with slope -5 and passes through point $(-2, -1)$? _____

21) The slope of a line is -10 and it passes through point $(-3, 0)$. What is the equation of the line? _____

22) The slope of a line is 8 and it passes through point $(0, 7)$. What is the equation of the line? _____

AFOQT Subject Test – Mathematics

Graphing Linear Inequalities

✏️ **Sketch the graph of each linear inequality.**

1) $y > 4x - 5$

2) $y < 2x + 4$

3) $y \leq -5x - 2$

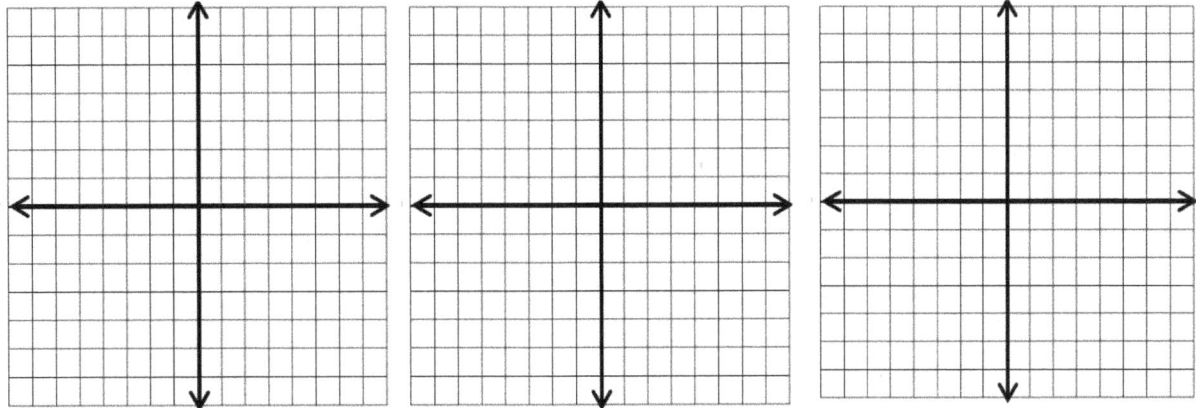

4) $4y \geq 12 + 4x$

5) $-12y < 3x - 24$

6) $5y \geq -15x + 10$

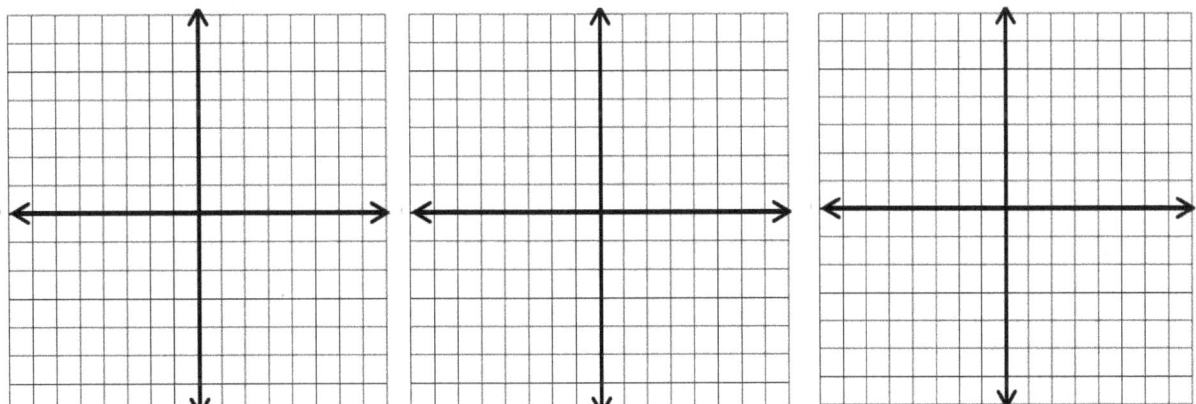

WWW.MathNotion.Com

AFOQT Subject Test – Mathematics

Finding Midpoint

✎ Find the midpoint of the line segment with the given endpoints.

1) $(-4, -3), (2, 3)$

2) $(9, 0), (-1, 8)$

3) $(9, -6), (3, 14)$

4) $(-10, -6), (0, 8)$

5) $(2, -5), (14, -15)$

6) $(-10, -3), (4, -13)$

7) $(8, 7), (-8, 13)$

8) $(-3, 6), (-9, 2)$

9) $(-4, 5), (16, -9)$

10) $(7, 14), (9, -2)$

11) $(-8, 6), (6, 6)$

12) $(10, 5), (-2, -3)$

13) $(-5, 12), (-3, 3)$

14) $(12, 7), (8, -2)$

15) $(10, 2), (-6, 14)$

16) $(-1, -2), (-7, 10)$

17) $(7, -7), (13, -13)$

18) $(-3, -8), (11, -4)$

19) $(5, -11), (-8, 9)$

20) $(14, -4), (16, 14)$

21) $(0, -5), (8, -1)$

22) $(3, 0), (-21, 18)$

23) $(17, -3), (-7, -5)$

24) $(26, -12), (6, 24)$

✎ Find the answer for each problem.

25) One endpoint of a line segment is $(-3, 7)$ and the midpoint of the line segment is $(-6, 9)$. What is the other endpoint? _____

26) One endpoint of a line segment is $(-3, 7)$ and the midpoint of the line segment is $(1, 5)$. What is the other endpoint? _____

27) One endpoint of a line segment is $(-10, -16)$ and the midpoint of the line segment is $(2, 9)$. What is the other endpoint? _____

AFOQT Subject Test – Mathematics

Finding Distance of Two Points

✎ **Find the distance between each pair of points.**

1) $(6, 3), (-3, -9)$

2) $(5, 2), (-10, -6)$

3) $(8, 5), (8, 3)$

4) $(-8, -2), (2, 22)$

5) $(6, -7), (-3, -7)$

6) $(12, 0), (-9, -20)$

7) $(3, 20), (3, -5)$

8) $(10, 17), (5, 5)$

9) $(7, -2), (-4, -2)$

10) $(13, 4), (5, -2)$

11) $(11, 13), (5, 5)$

12) $(1, 4), (-23, -3)$

13) $(9, 8), (5, -4)$

14) $(-11, -4), (5, 8)$

15) $(-2, -6), (-2, -12)$

16) $(-1, -4), (23, 3)$

17) $(19, 3), (7, -6)$

18) $(-5, -2), (3, 4)$

19) $(2, 6), (2, -12)$

20) $(-4, -2), (8, -2)$

✎ **Find the answer for each problem.**

21) Triangle ABC is a right triangle on the coordinate system and its vertices are $(-2, 5)$, $(-2, 1)$, and $(1, 1)$. What is the area of triangle ABC? _____

22) Three vertices of a triangle on a coordinate system are $(3, -6)$, $(-5, -12)$, and $(3, -18)$. What is the perimeter of the triangle? _____

23) Four vertices of a rectangle on a coordinate system are $(-2, 2)$, $(-2, 6)$, $(4, 2)$, and $(4, 6)$. What is its perimeter? _____

AFOQT Subject Test – Mathematics

Answers of Worksheets

Finding Slope

1) 1
2) −3
3) 2
4) −4
5) 6
6) −5
7) 8
8) −9
9) −7
10) 3
11) $\frac{1}{3}$
12) $-\frac{4}{5}$
13) $\frac{1}{2}$
14) −1
15) $\frac{1}{3}$
16) $\frac{1}{8}$
17) $\frac{7}{5}$
18) 2
19) −3
20) 2
21) −1
22) 2
23) 2
24) −3
25) −1
26) $\frac{5}{2}$
27) 2
28) −11

Graphing Lines Using Line Equation

1) $y = x - 2$

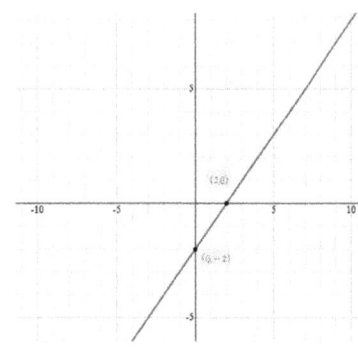

2) $y = -3x + 2$

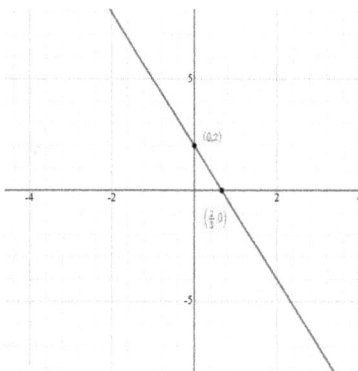

3) $x + y = 0$

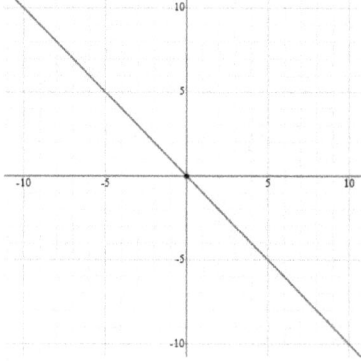

4) $x + y = -3$

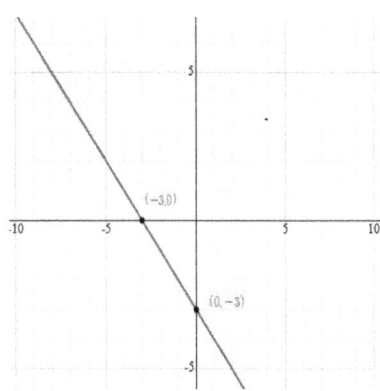

5) $2x + 3y = -4$

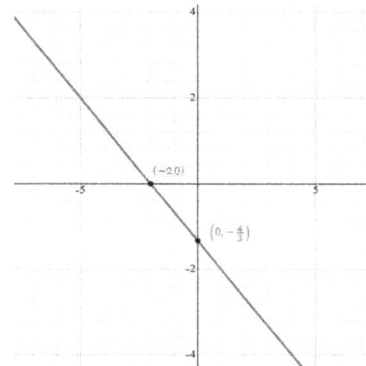

6) $y - 3x + 6 = 0$

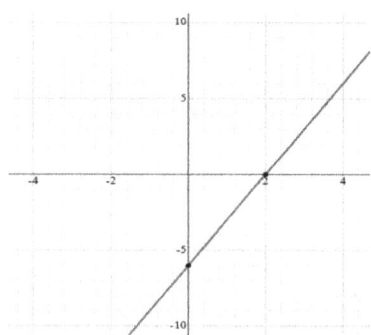

AFOQT Subject Test – Mathematics

Writing Linear Equations

1) $y = 14x - 33$
2) $y = x + 9$
3) $y = -4x + 47$
4) $y = x - 4$
5) $y = 2x - 31$
6) $y = 9x - 43$
7) $y = -2x + 24$
8) $y = -3x + 1$
9) $y = -2x + 5$
10) $y = -2x + 6$
11) $y = 5x - 18$
12) $y = 14x - 122$
13) $y = 13x + 52$
14) $y = x + 10$
15) $y = -5x + 58$
16) $y = -5x + 5$
17) $y = 6x + 12$
18) $y = -11x - 4$
19) $y = -3x + 17$
20) $y = -5x - 11$
21) $y = -10x - 30$
22) $y = 8x + 7$

Graphing Linear Inequalities

1) $y > 4x - 5$ 2) $y < 2x + 4$ 3) $y \leq -5x - 2$

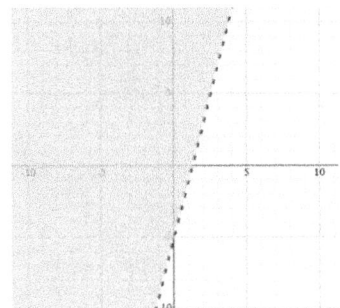   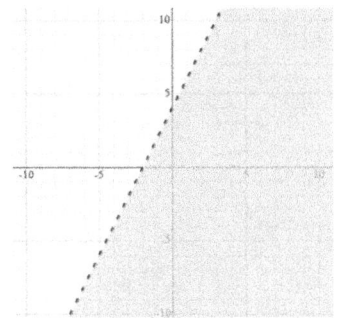   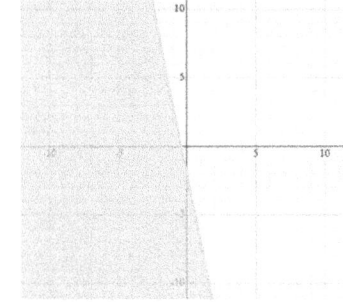

4) $4y \geq 12 + 4x$ 5) $-12y < 3x - 24$ 6) $5y \geq -15x + 10$

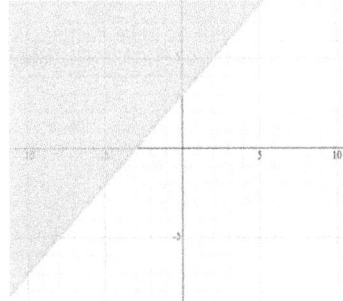   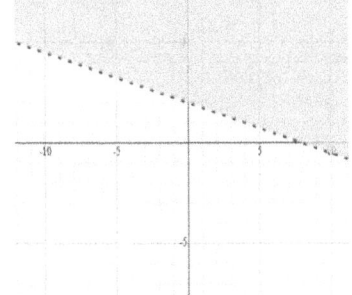   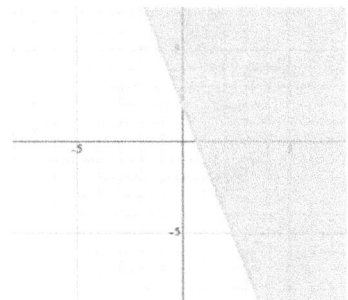

Finding Midpoint

1) $(-1, 0)$
2) $(4, 4)$
3) $(6, 4)$
4) $(-5, 1)$
5) $(8, -10)$
6) $(-3, -8)$
7) $(0, 10)$
8) $(-6, 4)$
9) $(6, -2)$
10) $(8, 6)$
11) $(-1, 6)$
12) $(4, 1)$
13) $(-4, 7.5)$
14) $(10, 2.5)$
15) $(2, 8)$
16) $(-4, 4)$
17) $(10, -10)$
18) $(4, -6)$

AFOQT Subject Test – Mathematics

19) $(-1.5, -1)$
20) $(15, 5)$
21) $(4, -3)$

22) $(-9, 9)$
23) $(5, -4)$
24) $(16, 6)$

25) $(-9, 11)$
26) $(5, 3)$
27) $(14, 34)$

Finding Distance of Two Points

1) 15
2) 17
3) 2
4) 26
5) 9
6) 29
7) 25
8) 13

9) 11
10) 10
11) 10
12) 25
13) $4\sqrt{10}$
14) 20
15) 6
16) 25

17) 15
18) 10
19) 18
20) 12
21) 6 square units
22) 32 units
23) 20 units

Chapter 8:
Polynomials

Topics that you'll practice in this chapter:

- ✓ Writing Polynomials in Standard Form
- ✓ Simplifying Polynomials
- ✓ Adding and Subtracting Polynomials
- ✓ Multiplying Monomials
- ✓ Multiplying and Dividing Monomials
- ✓ Multiplying a Polynomial and a Monomial
- ✓ Multiplying Binomials
- ✓ Factoring Trinomials
- ✓ Operations with Polynomials

Mathematics is the supreme judge; from its decisions there is no appeal. – Tobias Dantzig

Writing Polynomials in Standard Form

✎ Write each polynomial in standard form.

1) $11x - 7x =$

2) $-5 + 19x - 19x =$

3) $6x^5 - 12x^3 =$

4) $12 + 17x^4 - 12 =$

5) $5x^2 + 4x - 9x^3 =$

6) $-3x^2 + 12x^5 =$

7) $5x + 8x^3 - 2x^8 =$

8) $-7x^3 + 4x - 9x^6 =$

9) $3x^2 + 22 - 6x =$

10) $3 - 4x + 9x^4 =$

11) $13x^2 + 28x - 8x^3 =$

12) $16 + 4x^2 - 2x^3 =$

13) $19x^2 - 9x + 9x^4 =$

14) $3x^4 - 7x^2 - 2x^3 =$

15) $-51 + 3x^2 - 8x^4 =$

16) $7x^2 - 8x^6 + 4x^4 - 15 =$

17) $6x^4 - 4x^5 + 16 - 3x^3 =$

18) $-2x^6 + 4x - 7x^2 - 5x =$

19) $11x^7 + 8x^5 - 5x^7 - 3x^2 =$

20) $2x^2 - 12x^5 + 8x^2 + 3x^6 =$

21) $4x^5 - 11x^7 - 6x^3 + 16x^5 =$

22) $6x^3 + 3x^5 + 34x^4 - 8x^5 =$

23) $3x(4x + 5 - 2x^2) =$

24) $12x(x^6 + 4x^3) =$

25) $5x(3x^2 + 6x + 4) =$

26) $7x(4 - 2x + 6x^5) =$

27) $3x(4x^4 - 4x^3 + 2) =$

28) $4x(2x^5 + 6x^2 - 3) =$

29) $5x(3x^4 + 4x^3 + 2x) =$

30) $2x(3x - 2x^3 + 4x^6) =$

AFOQT Subject Test – Mathematics

Simplifying Polynomials

✎ **Simplify each expression.**

1) $3(4x - 20) =$

2) $5x(3x - 4) =$

3) $6x(5x - 7) =$

4) $3x(7x + 5) =$

5) $5x(4x - 3) =$

6) $6x(8x + 2) =$

7) $(3x - 2)(x - 4) =$

8) $(x - 5)(2x + 6) =$

9) $(x - 3)(x - 7) =$

10) $(3x + 4)(3x - 4) =$

11) $(5x - 4)(5x - 2) =$

12) $6x^2 + 6x^2 - 8x^4 =$

13) $3x - 2x^2 + 5x^3 + 7 =$

14) $7x + 4x^2 - 10x^3 =$

15) $12x^2 + 5x^5 - 6x^3 =$

16) $-5x^2 + 4x^6 + 6x^8 =$

17) $-12x^3 + 10x^5 - 4x^6 + 4x =$

18) $11 - 7x^2 + 4x^2 - 16x^3 + 11 =$

19) $2x^2 - 9x + 4x^3 + 15x - 10x =$

20) $13 - 7x^5 + 6x^5 - 4x^2 + 5 =$

21) $-5x^8 + x^6 - 14x^3 + 5x^8 =$

22) $(7x^4 - 4) + (7x^4 - 2x^4) =$

23) $3(3x^4 - 4x^3 - 6x^4) =$

24) $-5(x^9 + 8) - 5(10 - x^9) =$

25) $8x^3 - 9x^4 - 2x + 19 - 8x^3 =$

26) $11 - 8x^3 + 6x^3 - 7x^5 + 6 =$

27) $(5x^3 - 4x) - (6x - 2 - 6x^3) =$

28) $4x^2 - 5x^4 - x(3x^3 + 2x) =$

29) $6x + 6x^5 - 10 - 4(x^5 - 3) =$

30) $4 - 3x^4 + (6x^5 - 2x^4 + 5x^5) =$

31) $-(x^5 + 4) - 8(3 + x^5) =$

32) $(4x^3 - 3x) - (3x - 5x^3) =$

AFOQT Subject Test – Mathematics

Adding and Subtracting Polynomials

✎ **Add or subtract expressions.**

1) $(-2x^2 - 3) + (3x^2 + 4) =$

2) $(4x^3 + 6) - (7 - 2x^3) =$

3) $(4x^5 + 5x^2) - (2x^5 + 15) =$

4) $(6x^3 - 2x^2) + (5x^2 - 4x) =$

5) $(10x^4 + 28x) - (34x^4 + 6) =$

6) $(7x^2 - 3) + (7x^2 + 3) =$

7) $(9x^2 + 4) - (10 - 5x^2) =$

8) $(6x^2 + x^5) - (x^5 + 4) =$

9) $(4x^3 - x) + (3x - 7x^3) =$

10) $(11x + 10) - (8x + 10) =$

11) $(15x^3 - 3x) - (3x - 4x^3) =$

12) $(4x - x^5) - (6x^5 + 8x) =$

13) $(2x^2 - 7x^7) - (4x^7 - 6x) =$

14) $(3x^2 - 5) + (8x^2 + 4x^5) =$

15) $(9x^4 + 5x^5) - (x^5 - 9x^4) =$

16) $(-4x^3 - 2x) + (9x - 5x^3) =$

17) $(4x - 3x^2) - (148x^2 + x) =$

18) $(5x - 8x^4) - (3x^4 - 4x^2) =$

19) $(8x^4 - 4) + (2x^4 - 3x^2) =$

20) $(5x^6 + 7x^3) - (x^3 - 5x^6) =$

21) $(-2x^2 + 20x^5 + 5x^4) + (12x^4 + 8x^5 + 24x^2) =$

22) $(7x^4 - 9x^7 - 6x) - (-3x^4 - 9x^7 + 6x) =$

23) $(14x + 12x^4 - 18x^6) + (20x^4 + 18x^6 - 10x) =$

24) $(5x^8 - 6x^6 - 4x) - (5x^3 + 9x^6 - 7x) =$

25) $(11x^2 - 6x^4 - 3x) - (-4x^2 - 12x^4 + 9x) =$

26) $(-5x^9 + 14x^3 + 3x^7) + (10x^7 + 26x^3 + 3x^9) =$

WWW.MathNotion.Com

Multiplying Monomials

✏️ **Simplify each expression.**

1) $6u^8 \times (-u^2) =$

2) $(-5p^8) \times (-2p^3) =$

3) $4xy^3z^5 \times 3z^4 =$

4) $3u^5t \times 8ut^4 =$

5) $(-5a^2) \times (-7a^3b^6) =$

6) $-3a^4b^3 \times 6a^2b =$

7) $13xy^5 \times x^4y^4 =$

8) $6p^4q^3 \times (-8pq^6) =$

9) $8s^4t^3 \times 4st^3 =$

10) $(-6x^4y^3) \times 6x^2y =$

11) $3xy^7z \times 12z^3 =$

12) $24xy \times x^2y =$

13) $13pq^4 \times (-3p^2q) =$

14) $13s^3t^4 \times st^4 =$

15) $11p^5 \times (-6p^3) =$

16) $(-8p^3q^5r) \times 3pq^4r^6 =$

17) $(-4a^4) \times (-7a^3b) =$

18) $6u^6v^2 \times (-5u^3v^4) =$

19) $9u^5 \times (-3u) =$

20) $-6xy^5 \times 4x^2y =$

21) $13y^5z^3 \times (-y^3z) =$

22) $8a^4bc^3 \times 2abc^3 =$

23) $(-7p^5q^6) \times (-5p^4q^2) =$

24) $4u^5v^3 \times (-4u^7v^3) =$

25) $17y^4z^5 \times (-y^6z) =$

26) $(-5pq^3r^2) \times 8p^2q^4r =$

27) $3ab^5c^6 \times 5a^4bc^2 =$

28) $6x^3yz^2 \times 3x^2y^7z^3 =$

AFOQT Subject Test – Mathematics

Multiplying and Dividing Monomials

✏️ **Simplify each expression.**

1) $(5x^5)(2x^2) =$

2) $(4x^4)(6x^2) =$

3) $(3x^4)(7x^4) =$

4) $(5x^6)(4x^2) =$

5) $(12x^4)(3x^6) =$

6) $(4yx^8)(8y^4x^3) =$

7) $(14x^4y)(x^3y^5) =$

8) $(-5x^3y^4)(2x^3y^5) =$

9) $(-6x^4y^2)(-3x^3y^5) =$

10) $(5x^3y)(-5x^2y^3) =$

11) $(6x^4y^3)(4x^3y^4) =$

12) $(4x^3y^2)(5x^2y^4) =$

13) $(12x^3y^6)(4x^4y^{10}) =$

14) $(15x^3y^5)(3x^4y^6) =$

15) $(7x^2y^7)(8x^6y^7) =$

16) $(-3x^3y^8)(7x^9y^4) =$

17) $\dfrac{5x^6y^6}{xy^4} =$

18) $\dfrac{19x^7y^5}{19x^6y} =$

19) $\dfrac{56x^4y^4}{8xy} =$

20) $\dfrac{81x^5y^6}{9x^4y^5} =$

21) $\dfrac{36x^7y^6}{9x^2y^3} =$

22) $\dfrac{48x^9y^7}{4x^4y^6} =$

23) $\dfrac{88x^{18}y^{12}}{11x^8y^9} =$

24) $\dfrac{30x^7y^6}{6x^8y^3} =$

25) $\dfrac{150x^7y^6}{30x^4y^6} =$

26) $\dfrac{-42x^{18}y^{14}}{6x^4y^9} =$

27) $\dfrac{-36x^7y^8}{9x^5y^8} =$

WWW.MathNotion.Com

AFOQT Subject Test – Mathematics

Multiplying a Polynomial and a Monomial

✎ Find each product.

1) $x(2x + 4) =$

2) $6(4 - 2x) =$

3) $5x(4x + 2) =$

4) $x(-4x + 5) =$

5) $8x(2x - 2) =$

6) $6(2x - 4y) =$

7) $7x(5x - 5) =$

8) $3x(12x + 2y) =$

9) $4x(x + 6y) =$

10) $11x(3x + 4y) =$

11) $7x(3x + 2) =$

12) $10x(4x - 10y) =$

13) $9x(3x - 2y) =$

14) $7x(x - 4y + 6) =$

15) $8x(2x^2 + 5y^2) =$

16) $12x(2x + 3y) =$

17) $4(2x^4 - 4y^4) =$

18) $4x(-3x^2y + 4y) =$

19) $-4(5x^3 - 2xy + 4) =$

20) $4(x^2 - 5xy - 6) =$

21) $8x(2x^3 - 5xy + 2x) =$

22) $-6x(-2x^3 - 6x + 2xy) =$

23) $3(2x^2 + xy - 9y^2) =$

24) $4x(5x^3 - 3x + 7) =$

25) $6(3x^{22} - 2x - 5) =$

26) $x^2(-2x^3 + 4x + 3) =$

27) $x^2(4x^3 + 10 - 2x) =$

28) $4x^4(3x^3 - 2x + 5) =$

29) $2x^2(4x^4 - 5xy + 7y^3) =$

30) $5x^2(5x^4 - 3x + 9) =$

31) $7x^2(6x^2 + 3x - 6) =$

32) $4x(x^3 - 4xy + 2y^2) =$

AFOQT Subject Test – Mathematics

Multiplying Binomials

✎ **Find each product.**

1) $(x + 3)(x + 6) =$

2) $(x - 4)(x + 3) =$

3) $(x - 3)(x - 8) =$

4) $(x + 8)(x + 9) =$

5) $(x - 2)(x - 12) =$

6) $(x + 5)(x + 5) =$

7) $(x - 6)(x + 7) =$

8) $(x - 8)(x - 3) =$

9) $(x + 7)(x + 12) =$

10) $(x - 4)(x + 8) =$

11) $(x + 8)(x + 8) =$

12) $(x + 2)(x + 7) =$

13) $(x - 6)(x + 6) =$

14) $(x - 5)(x + 5) =$

15) $(x + 11)(x + 11) =$

16) $(x + 6)(x + 9) =$

17) $(x - 2)(x + 2) =$

18) $(x - 4)(x + 7) =$

19) $(3x + 5)(x + 6) =$

20) $(5x - 6)(4x + 8) =$

21) $(x - 7)(3x + 7) =$

22) $(x - 9)(x - 4) =$

23) $(x - 12)(x + 2) =$

24) $(2x - 4)(5x + 4) =$

25) $(3x - 8)(x + 8) =$

26) $(7x - 2)(6x + 3) =$

27) $(4x + 5)(3x + 5) =$

28) $(7x - 4)(9x + 4) =$

29) $(x + 2)(2x - 8) =$

30) $(5x - 4)(5x + 4) =$

31) $(3x + 2)(3x - 7) =$

32) $(x^2 + 8)(x^2 - 8) =$

WWW.MathNotion.Com

AFOQT Subject Test – Mathematics

Factoring Trinomials

✎ **Factor each trinomial.**

1) $x^2 + 8x + 12 =$

2) $x^2 - 6x + 5 =$

3) $x^2 + 15x + 36 =$

4) $x^2 - 12x + 35 =$

5) $x^2 - 11x + 18 =$

6) $x^2 - 9x + 18 =$

7) $x^2 + 18x + 72 =$

8) $x^2 - x - 72 =$

9) $x^2 + 4x - 21 =$

10) $x^2 - 13x + 22 =$

11) $x^2 + 2x - 24 =$

12) $x^2 - 3x - 40 =$

13) $x^2 - 3x - 70 =$

14) $x^2 + 26x + 169 =$

15) $4x^2 - 7x - 15 =$

16) $x^2 - 14x + 33 =$

17) $10x^2 + 5x - 15 =$

18) $6x^2 - 4x - 42 =$

19) $x^2 + 12x + 36 =$

20) $5x^2 + 17x - 12 =$

✎ **Calculate each problem.**

21) The area of a rectangle is $x^2 - x - 56$. If the width of rectangle is $x + 7$, what is its length? _____

22) The area of a parallelogram is $4x^2 + 17x - 15$ and its height is $x + 5$. What is the base of the parallelogram? _____

23) The area of a rectangle is $6x^2 - 22x + 12$. If the width of the rectangle is $3x - 2$, what is its length? _____

AFOQT Subject Test – Mathematics

Operations with Polynomials

✏️ **Find each product.**

1) $4(5x + 3) = $ _____

2) $8(2x + 6) = $ _____

3) $2(5x - 2) = $ _____

4) $-4(7x - 3) = $ _____

5) $3x^2(9x + 1) = $ _____

6) $4x^6(7x - 9) = $ _____

7) $3x^4(-7x + 3) = $ _____

8) $-8x^4(5x - 8) = $ _____

9) $7(x^2 + 5x - 3) = $ _____

10) $9(5x^2 - 7x + 5) = $ _____

11) $3(3x^2 + 3x + 2) = $ _____

12) $5x(3x^2 + 5x + 8) = $ _____

13) $(5x + 7)(3x - 3) = $ _____

14) $(9x + 3)(3x - 5) = $ _____

15) $(6x + 3)(4x - 2) = $ _____

16) $(7x - 2)(3x + 5) = $ _____

✏️ **Calculate each problem.**

17) The measures of two sides of a triangle are $(2x + 5y)$ and $(6x - 3y)$. If the perimeter of the triangle is $(13x + 4y)$, what is the measure of the third side? _____

18) The height of a triangle is $(8x + 5)$ and its base is $(4x - 3)$. What is the area of the triangle? _____

19) One side of a square is $(6x + 2)$. What is the area of the square? _____

20) The length of a rectangle is $(5x - 8y)$ and its width is $(15x + 8y)$. What is the perimeter of the rectangle? _____

21) The side of a cube measures $(x + 2)$. What is the volume of the cube? _____

22) If the perimeter of a rectangle is $(28x + 6y)$ and its width is $(5x + 2y)$, what is the length of the rectangle? _____

AFOQT Subject Test – Mathematics

Answers of Worksheets

Writing Polynomials in Standard Form

1) $4x$
2) -5
3) $6x^5 - 12x^3$
4) $14x^4$
5) $-9x^3 + 5x^2 + 4x$
6) $12x^5 - 3x^2$
7) $-2x^8 + 8x^3 + 5x$
8) $-9x^6 - 7x^3 + 4x$
9) $3x^2 - 6x + 22$
10) $9x^4 - 4x + 3$
11) $-8x^3 + 13x^2 + 28x$
12) $-2x^3 + 4x^2 + 16$
13) $9x^4 + 19x^2 - 9x$
14) $3x^4 - 2x^3 - 7x^2$
15) $-8x^4 + 3x^2 - 51$
16) $-8x^6 + 4x^4 + 7x^2 - 15$
17) $-4x^5 + 6x^4 - 3x^3 + 16$
18) $-2x^6 - 7x^2 - x$
19) $6x^7 + 8x^5 - 3x^2$
20) $3x^6 - 12x^5 + 10x^2$
21) $-11x^7 + 20x^5 - 6x^3$
22) $-5x^5 + 34x^4 + 6x^3$
23) $-6x^3 + 12x^2 + 15x$
24) $12x^7 + 48x^4$
25) $15x^3 + 30x^2 + 20x$
26) $42x^6 - 14x^2 + 28x$
27) $12x^5 - 12x^4 + 6x$
28) $8x^6 + 24x^3 - 12x$
29) $15x^5 + 20x^4 + 10x^2$
30) $8x^7 - 4x^4 + 6x^2$

Simplifying Polynomials

1) $12x - 60$
2) $15x^2 - 20x$
3) $30x^2 - 42x$
4) $21x^2 + 15x$
5) $20x^2 - 15x$
6) $48x^2 + 12x$
7) $3x^2 - 14x + 8$
8) $2x^2 - 4x - 30$
9) $x^2 - 10x + 21$
10) $9x^2 - 16$
11) $25x^2 - 30x + 8$
12) $-8x^4 + 12x^2$
13) $5x^3 - 2x^2 + 3x + 7$
14) $-10x^3 + 4x^2 + 7x$
15) $5x^5 - 6x^3 + 12x^2$
16) $6x^8 + 4x^6 - 5x^2$
17) $-4x^6 + 10x^5 - 12x^3 + 4x$
18) $-16x^3 - 3x^2 + 22$
19) $4x^3 + 2x^2 - 4x$
20) $-x^5 - 4x^2 + 18$
21) $x^6 - 14x^3$
22) $12x^4 - 4$
23) $-9x^4 - 12x^3$
24) -90

WWW.MathNotion.Com

AFOQT Subject Test – Mathematics

25) $-9x^4 - 2x + 19$

26) $-7x^5 - 2x^3 + 17$

27) $11x^3 - 10x + 2$

28) $-8x^4 + 2x^2$

29) $2x^5 + 6x + 2$

30) $11x^5 - 5x^4 + 4$

31) $-9x^5 - 28$

32) $9x^3 - 6x$

Adding and Subtracting Polynomials

1) $x^2 + 1$

2) $6x^3 - 1$

3) $2x^5 + 5x^2 - 15$

4) $6x^3 + 3x^2 - 4x$

5) $-24x^4 + 28x - 6$

6) $14x^2$

7) $14x^2 - 6$

8) $6x^2 - 4$

9) $-3x^3 + 2x$

10) $3x$

11) $19x^3 - 6x$

12) $-7x^5 - 4x$

13) $-11x^7 + 2x^2 + 6x$

14) $4x^5 + 11x^2 - 5$

15) $4x^5 + 18x^4$

16) $-9x^3 + 7x$

17) $-151x^2 + 3x$

18) $-11x^4 + 4x^2 + 5x$

19) $10x^4 - 3x^2 - 4$

20) $10x^6 + 6x^3$

21) $28x^5 + 17x^4 + 22x^2$

22) $10x^4 - 12x$

23) $32x^4 + 4x$

24) $5x^8 - 15x^6 - 5x^3 + 3x$

25) $6x^4 + 15x^2 - 12x$

26) $-2x^9 + 13x^7 + 40x^3$

Multiplying Monomials

1) $-6u^{10}$

2) $10p^{11}$

3) $12xy^3z^9$

4) $24u^6t^5$

5) $35a^5b^6$

6) $-18a^6b^4$

7) $13x^5y^9$

8) $-48p^5q^9$

9) $32s^5t^6$

10) $-36x^6y^4$

11) $36xy^7z^4$

12) $24px^3y^2$

13) $-39p^3q^5$

14) $13s^4t^8$

15) $-66p^8$

16) $-24p^4q^9r^7$

17) $28a^7b$

18) $-30u^9v^6$

19) $-27u^6$

20) $-24x^3y^6$

21) $-13y^8z^4$

22) $16a^5b^2c^6$

23) $35p^9q^8$

24) $-16u^{12}v^6$

25) $-17y^{10}z^6$

26) $-40p^3q^7r^3$

27) $15a^5b^6c^8$

28) $18x^5y^8z^5$

Multiplying and Dividing Monomials

1) $10x^7$

2) $24x^6$

3) $21x^8$

4) $20x^8$

5) $36x^{10}$

6) $32x^{11}y^5$

7) $14x^7y^6$

8) $-10x^6y^9$

9) $18x^7y^7$

10) $-25x^5y^4$

11) $24x^7y^7$

12) $20x^5y^6$

AFOQT Subject Test – Mathematics

13) $48x^7y^{16}$
14) $45x^7y^{11}$
15) $56x^8y^{14}$
16) $-21x^{12}y^{12}$
17) $5x^5y^2$

18) xy^4
19) $7x^3y^3$
20) $9xy$
21) $4x^5y^3$
22) $12x^5y$

23) $8x^{10}y^3$
24) $5x^{-1}y^3$
25) $5x^3$
26) $-7x^{14}y^5$
27) $-4x^2$

Multiplying a Polynomial and a Monomial

1) $2x^2 + 4x$
2) $-12x + 24$
3) $20x^2 + 10x$
4) $-4x^2 + 5x$
5) $16x^2 - 16x$
6) $12x - 24y$
7) $35x^2 - 35x$
8) $36x^2 + 6xy$
9) $4x^2 + 24xy$
10) $33x^2 + 44xy$
11) $21x^2 + 14x$
12) $40x^2 - 100xy$
13) $27x^2 - 18xy$
14) $7x^2 - 28xy + 42x$
15) $16x^3 + 40xy^2$
16) $24x^2 + 36xy$

17) $8x^4 - 16y^4$
18) $-12x^3y + 16xy$
19) $-20x^3 + 8xy - 16$
20) $4x^2 - 20xy - 24$
21) $16x^4 - 40x^2y + 16x^2$
22) $12x^4 + 36x^2 - 12x^2y$
23) $6x^2 + 3xy - 27y^2$
24) $20x^4 - 12x^2 + 28x$
25) $18x^{22} - 12x - 30$
26) $-2x^5 + 4x^3 + 3x^2$
27) $4x^5 - 2x^3 + 10x^2$
28) $12x^7 - 8x^5 + 20x^4$
29) $8x^6 - 10x^3y + 14x^2y^3$
30) $25x^6 - 15x^3 + 45x^2$
31) $42x^4 + 21x^3 - 42x^2$
32) $4x^4 - 16x^2y + 8xy^2$

Multiplying Binomials

1) $x^2 + 9x + 18$
2) $x^2 - x - 12$
3) $x^2 - 11x + 24$
4) $x^2 + 17x + 72$
5) $x^2 - 14x + 24$
6) $x^2 + 10x + 25$
7) $x^2 + x - 42$

8) $x^2 - 11x + 24$
9) $x^2 + 19x + 84$
10) $x^2 + 4x - 32$
11) $x^2 + 16x + 64$
12) $x^2 + 9x + 14$
13) $x^2 - 36$
14) $x^2 - 25$

WWW.MathNotion.Com

AFOQT Subject Test – Mathematics

15) $x^2 + 22x + 121$
16) $x^2 + 15x + 54$
17) $x^2 - 4$
18) $x^2 + 3x - 28$
19) $3x^2 + 23x + 30$
20) $20x^2 + 16x - 48$
21) $3x^2 - 14x - 49$
22) $x^2 - 13x + 36$
23) $x^2 - 10x - 24$

24) $10x^2 - 12x - 16$
25) $3x^2 + 16x - 64$
26) $42x^2 + 9x - 6$
27) $12x^2 + 35x + 25$
28) $63x^2 - 8x - 16$
29) $2x^2 - 4x - 16$
30) $25x^2 - 16$
31) $9x^2 - 15x - 14$
32) $x^4 - 64$

Factoring Trinomials

1) $(x + 6)(x + 2)$
2) $(x - 5)(x - 1)$
3) $(x + 12)(x + 3)$
4) $(x - 5)(x - 7)$
5) $(x - 2)(x - 9)$
6) $(x - 6)(x - 3)$
7) $(x + 6)(x + 12)$
8) $(x + 8)(x - 9)$

9) $(x - 3)(x + 7)$
10) $(x - 11)(x - 2)$
11) $(x - 4)(x + 6)$
12) $(x - 8)(x + 5)$
13) $(x + 7)(x - 10)$
14) $(x + 13)(x + 13)$
15) $(4x + 5)(x - 3)$
16) $(x - 11)(x - 3)$

17) $(5x - 5)(2x + 3)$
18) $(2x - 6)(3x + 7)$
19) $(x + 6)(x + 6)$
20) $(5x - 3)(x + 4)$
21) $(x - 8)$
22) $(4x - 3)$
23) $(2x - 6)$

Operations with Polynomials

1) $20x + 12$
2) $16x + 48$
3) $10x - 4$
4) $-28x + 12$
5) $27x^3 + 3x^2$
6) $28x^7 - 36x^6$
7) $-21x^5 + 9x^4$
8) $-40x^5 + 64x^4$

9) $7x^2 + 35x - 21$
10) $45x^2 - 63x + 45$
11) $9x^2 + 9x + 6$
12) $15x^3 + 25x^2 + 40x$
13) $15x^2 + 6x - 21$
14) $27x^2 - 36x - 15$
15) $24x^2 - 6$
16) $21x^2 + 29x - 10$

17) $(5x + 2y)$
18) $16x^2 - 2x - \frac{15}{2}$
19) $36x^2 + 24x + 4$
20) $40x$
21) $x^3 + 6x^2 + 12x + 8$
22) $(9x + y)$

WWW.MathNotion.Com

Chapter 9 : Functions Operations and Quadratic

Topics that you'll practice in this chapter:

- ✓ Evaluating Function
- ✓ Adding and Subtracting Functions
- ✓ Multiplying and Dividing Functions
- ✓ Composition of Functions
- ✓ Quadratic Equation
- ✓ Solving Quadratic Equations
- ✓ Quadratic Formula and the Discriminant
- ✓ Graphing Quadratic Functions

It's fine to work on any problem, so long as it generates interesting mathematics along the way – even if you don't solve it at the end of the day." – Andrew Wiles

AFOQT Subject Test – Mathematics

Evaluating Function

✍ **Write each of following in function notation.**

1) $h = -8x + 3$

2) $k = 2a - 14$

3) $d = 11t$

4) $y = \frac{5}{12}x - \frac{7}{12}$

5) $m = 24n - 210$

6) $c = p^2 - 5p + 10$

✍ **Evaluate each function.**

7) $f(x) = 2x - 7$, find $f(-3)$

8) $g(x) = \frac{1}{9}x + 12$, find $f(18)$

9) $h(x) = -4x + 9$, find $f(3)$

10) $f(x) = -x + 19$, find $f(-3)$

11) $f(a) = 7a - 12$, find $f(3)$

12) $h(x) = 14 - 3x$, find $f(-4)$

13) $g(n) = 6n - 10$, find $f(2)$

14) $f(x) = -11x - 4$, find $f(-1)$

15) $k(n) = -20 - 3.5n$, find $f(2)$

16) $f(x) = -0.7x + 3.3$, find $f(-7)$

17) $g(n) = \frac{11n+8}{n}$, find $g(2)$

18) $g(n) = \sqrt{3n} + 12$, find $g(3)$

19) $h(x) = x^{-2} - 7$, find $h(\frac{1}{9})$

20) $h(n) = n^{-3} + 11$, find $h(\frac{1}{4})$

21) $h(n) = n^3 - 2$, find $h(\frac{1}{2})$

22) $h(n) = n^2 - 4$, find $h(-\frac{1}{3})$

23) $h(n) = 4n^2 - 13$, find $h(-5)$

24) $h(n) = -2n^3 - 6n$, find $h(2)$

25) $g(n) = \sqrt{16n^2} - \sqrt{n}$, find $g(4)$

26) $h(a) = \frac{-14a+9}{3a}$, find $h(-b)$

27) $k(a) = 12a - 14$, find $k(a - 3)$

28) $h(x) = \frac{1}{9}x + 18$, find $h(-18x)$

29) $h(x) = 8x^2 + 16$, find $h(\frac{x}{2})$

30) $h(x) = x^4 - 20$, find $h(-2x)$

WWW.MathNotion.Com

AFOQT Subject Test – Mathematics

Adding and Subtracting Functions

✎ **Perform the indicated operation.**

1) $f(x) = 2x + 3$

   $g(x) = x + 7$

   Find $(f - g)(2)$

2) $g(a) = -5a - 8$

   $f(a) = -3a - 5$

   Find $(g - f)(-2)$

3) $h(t) = 4t + 3$

   $g(t) = 4t + 7$

   Find $(h - g)(t)$

4) $g(a) = -6a - 10$

   $f(a) = 3a^2 + 9$

   Find $(g - f)(x)$

5) $g(x) = \frac{5}{6}x - 23$

   $h(x) = \frac{5}{12}x + 25$

   Find $g(12) - h(12)$

6) $h(x) = \sqrt{3x} - 2$

   $g(x) = \sqrt{3x} + 5$

   Find $(h + g)(12)$

7) $f(x) = x^{-1}$

   $g(x) = x^2 + \frac{5}{x}$

   Find $(f - g)(-3)$

8) $h(n) = n^2 + 2$

   $g(n) = -4n + 6$

   Find $(h - g)(2a)$

9) $g(x) = -2x^2 - 5 - 4x$

   $f(x) = 7 + 2x$

   Find $(g - f)(3x)$

10) $g(t) = 11t - 4$

    $f(t) = -2t^2 + 5$

    Find $(g + f)(-t)$

11) $f(x) = 8x + 9$

    $g(x) = -5x^2 + 3x$

    Find $(f - g)(-x^2)$

12) $f(x) = -3x^4 - 5x$

    $g(x) = 2x^4 + 5x + 22$

    Find $(f + g)(3x^2)$

AFOQT Subject Test – Mathematics

Multiplying and Dividing Functions

✎ **Perform the indicated operation.**

1) $g(x) = -2x - 1$
   $f(x) = 4x + 3$
   Find $(g.f)(2)$

2) $f(x) = 5x$
   $h(x) = -2x + 3$
   Find $(f.h)(-2)$

3) $g(a) = 5a - 2$
   $h(a) = 2a - 3$
   Find $(g.h)(-3)$

4) $f(x) = 2x - 7$
   $h(x) = x - 5$
   Find $(\frac{f}{h})(4)$

5) $f(x) = 8a^2$
   $g(x) = 3 + 2a$
   Find $(\frac{f}{g})(2)$

6) $g(a) = \sqrt{4a} + 2$
   $f(a) = (-a)^4 + 1$
   Find $(\frac{g}{f})(1)$

7) $g(t) = t^3 + 1$
   $h(t) = 5t - 2$
   Find $(g.h)(-2)$

8) $g(n) = n^2 + 2n - 4$
   $h(n) = -5n + 3$
   Find $(g.h)(1)$

9) $g(a) = (a - 3)^2$
   $f(a) = a^2 + 4$
   Find $(\frac{g}{f})(3)$

10) $g(x) = -3x^2 + \frac{4}{5}x + 9$
    $f(x) = x^2 - 24$
    Find $(\frac{g}{f})(5)$

11) $f(x) = 2x^3 - 5x^2 + 1$
    $g(x) = 3x - 1$
    Find $(f.g)(x)$

12) $f(x) = 5x - 2$
    $g(x) = x^3 - 2x$
    Find $(f.g)(x^2)$

WWW.MathNotion.Com

AFOQT Subject Test – Mathematics

Composition of Functions

Using $f(x) = 2x - 5$ and $g(x) = -2x$, find:

1) $f(g(2)) =$

2) $f(g(-1)) =$

3) $g(f(-4)) =$

4) $g(f(5)) =$

5) $f(g(3)) =$

6) $g(f(0)) =$

Using $f(x) = -\frac{1}{4}x + \frac{3}{4}$ and $g(x) = 2x^2$, find:

7) $g(f(-2)) =$

8) $g(f(4)) =$

9) $g(g(1)) =$

10) $f(f(1)) =$

11) $g(f(-4)) =$

12) $f(g(x)) =$

Using $f(x) = -2x + 2$ and $g(x) = x + 1$, find:

13) $g(f(1)) =$

14) $f(f(0)) =$

15) $f(g(-1)) =$

16) $f(g(-3)) =$

17) $g(f(2)) =$

18) $f(g(x)) =$

Using $f(x) = \sqrt{x + 9}$ and $g(x) = x - 9$, find:

19) $f(g(9)) =$

20) $g(f(-9)) =$

21) $f(g(4)) =$

22) $f(f(7)) =$

23) $g(f(-5)) =$

24) $g(g(0)) =$

Quadratic Equation

✏ Multiply.

1) $(x-4)(x+6) = $ _____

2) $(x+5)(x+7) = $ _____

3) $(x-6)(x+8) = $ _____

4) $(x+2)(x-9) = $ _____

5) $(x-7)(x-8) = $ _____

6) $(3x+2)(x-3) = $ _____

7) $(4x-3)(x+2) = $ _____

8) $(4x-5)(x+1) = $ _____

9) $(7x+1)(x-6) = $ _____

10) $(5x+1)(3x-3) = $ _____

✏ Factor each expression.

11) $x^2 - 2x - 8 = $ _____

12) $x^2 + 8x + 15 = $ _____

13) $x^2 - 2x - 24 = $ _____

14) $x^2 - 10x + 21 = $ _____

15) $x^2 + 10x + 21 = $ _____

16) $4x^2 + 9x + 5 = $ _____

17) $5x^2 + 13x - 6 = $ _____

18) $5x^2 + 17x - 12 = $ _____

19) $2x^2 + 7x + 5 = $ _____

20) $9x^2 - 21x + 6 = $ _____

✏ Calculate each equation.

21) $(x+6)(x-3) = 0$

22) $(x+1)(x+8) = 0$

23) $(3x+6)(x+5) = 0$

24) $(2x-2)(4x+8) = 0$

25) $x^2 + x + 10 = 22$

26) $x^2 + 11x + 36 = 12$

27) $2x^2 + 9x + 9 = 5$

28) $x^2 + 3x - 24 = 4$

29) $5x^2 + 5x - 40 = 20$

30) $8x^2 + 8x = 48$

AFOQT Subject Test – Mathematics

Solving Quadratic Equations

✎ **Solve each equation by factoring or using the quadratic formula.**

1) $(x+9)(x-1) = 0$

2) $(x+7)(x+6) = 0$

3) $(x-8)(x+3) = 0$

4) $(x-6)(x-4) = 0$

5) $(x+2)(x+12) = 0$

6) $(5x+4)(x+7) = 0$

7) $(6x+1)(4x+5) = 0$

8) $(2x+7)(x+8) = 0$

9) $(x+6)(3x+15) = 0$

10) $(12x+2)(x+8) = 0$

11) $x^2 = 8x$

12) $x^2 - 16 = 0$

13) $3x^2 + 6 = 9x$

14) $-2x^2 - 8 = 10x$

15) $5x^2 + 40x = 45$

16) $x^2 + 10x = 24$

17) $x^2 + 6x = 16$

18) $x^2 + 9x = -18$

19) $x^2 + 13x = -36$

20) $x^2 + 3x - 15 = 5x$

21) $x^2 + 8x + 7 = -8$

22) $3x^2 - 11x = -9 + x$

23) $10x^2 + 3 = 27x - 15$

24) $7x^2 - 6x + 8 = 8$

25) $2x^2 - 12 = -3x + 2$

26) $10x^2 - 26x - 3 = -15$

27) $3x^2 + 21 = -16x + 5$

28) $x^2 + 15x - 10 = -66$

29) $3x^2 - 8x - 8 = 4 + x$

30) $2x^2 + 6x - 24 = 12$

31) $3x^2 - 33x + 54 = -18$

32) $-10x^2 - 15x - 9 = -9 - 27x^2$

WWW.MathNotion.Com

AFOQT Subject Test – Mathematics

Quadratic Formula and the Discriminant

✎ Find the value of the discriminant of each quadratic equation.

1) $3x(x - 8) = 0$

2) $2x^2 + 6x - 4 = 0$

3) $x^2 + 6x + 7 = 0$

4) $x^2 - x + 3 = 0$

5) $x^2 + 4x - 3 = 0$

6) $2x^2 + 6x - 10 = 0$

7) $3x^2 + 7x + 5 = 0$

8) $x^2 - 6x - 4 = 0$

9) $2x^2 + 8x + 3 = 0$

10) $x^2 + 7x - 5 = 0$

11) $5x^2 + 2x - 3 = 0$

12) $-3x^2 - 11x + 4 = 0$

13) $-6x^2 - 12x + 8 = 0$

14) $-x^2 - 9x - 12 = 0$

15) $7x^2 - 6x - 10 = 0$

16) $-4x^2 - 2x + 8 = 0$

17) $5x^2 + 8x - 2 = 0$

18) $6x^2 - 4x = 0$

19) $3x^2 - 5x + 2 = 0$

20) $4x^2 + 9x + 3 = 0$

✎ Find the discriminant of each quadratic equation then state the number of real and imaginary solutions.

21) $-4x^2 - 16 = 16x$

22) $20x^2 = 20x - 5$

23) $-11x^2 - 19x = 26$

24) $22x^2 - 4x + 1 = 18x^2$

25) $-11x^2 = -15x + 8$

26) $3x^2 + 6x + 9 = 6$

27) $13x^2 - 5x - 12 = -26$

28) $-8x^2 - 32x - 25 = 7$

WWW.MathNotion.Com

AFOQT Subject Test – Mathematics

Graphing Quadratic Functions

✎ Sketch the graph of each function. Identify the vertex and axis of symmetry.

1) $y = (x + 3)^2 + 2$

2) $y = (x - 3)^2 - 2$

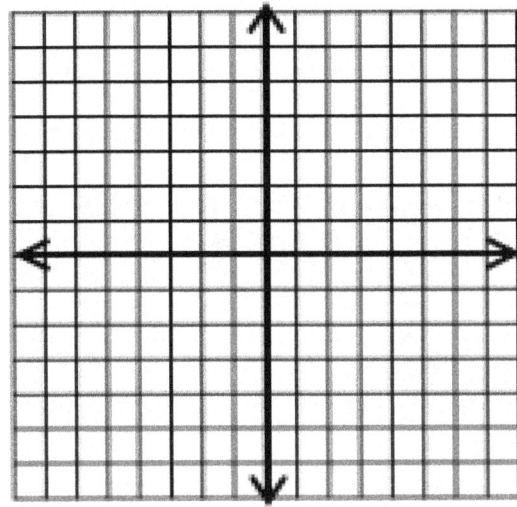
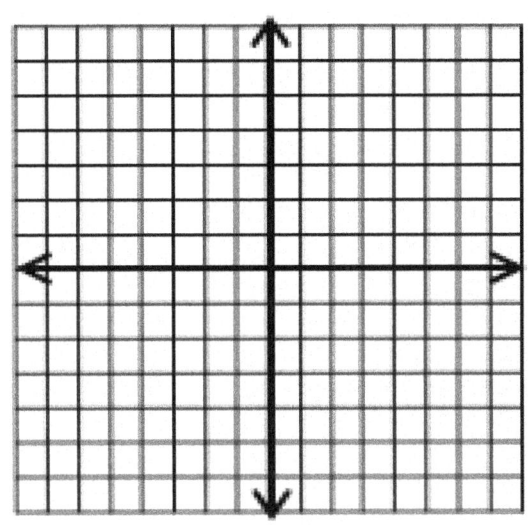

3) $y = 6 - (-x + 4)^2$

4) $y = -3x^2 - 6x + 9$

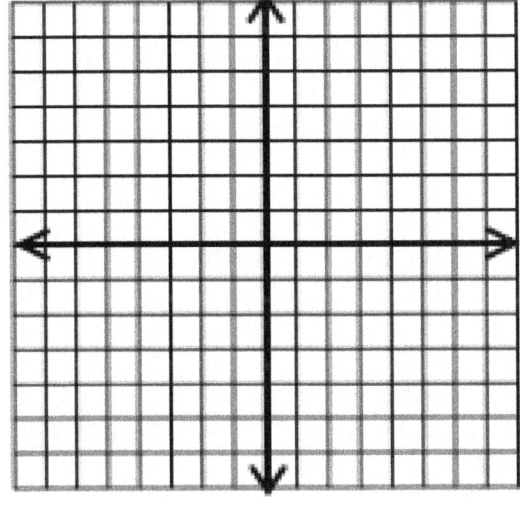
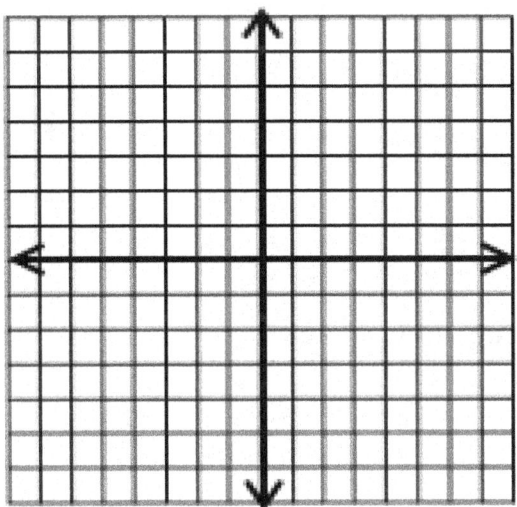

WWW.MathNotion.Com

AFOQT Subject Test – Mathematics

Answers of Worksheets

Evaluating Function

1) $h(x) = -8x + 3$
2) $k(a) = 2a - 14$
3) $d(t) = 11t$
4) $f(x) = \frac{5}{12}x - \frac{7}{12}$
5) $m(n) = 24n - 210$
6) $c(p) = p^2 - 5p + 10$
7) -13
8) 14
9) -3
10) 22
11) 9
12) 26
13) 2
14) 7
15) -27
16) 8.2
17) 15
18) 15
19) 74
20) 75
21) $-1\frac{7}{8}$
22) $-3\frac{8}{9}$
23) 87
24) -28
25) 14
26) $-\frac{14b+9}{3b}$
27) $12a - 50$
28) $-2x + 18$
29) $2x^2 + 16$
30) $16x^4 - 20$

Adding and Subtracting Functions

1) -2
2) 1
3) -4
4) $-3x^2 - 6x - 19$
5) -43
6) 15
7) $-7\frac{2}{3}$
8) $4a^2 + 8a - 4$
9) $-18x^2 - 18x - 12$
10) $-2t^2 - 11t + 1$
11) $5x^4 - 5x^2 + 9$
12) $-81x^8 + 22$

Multiplying and Dividing Functions

1) -55
2) -70
3) 153
4) -1
5) $4\frac{4}{7}$
6) 2
7) 84
8) 2
9) 0
10) -62
11) $6x^4 - 17x^3 + 5x^2 + 3x - 1$
12) $5x^8 - 2x^6 - 10x^4 + 4x^2$

Composition of Functions

1) -13
2) -1
3) 26
4) -10
5) -17
6) 10
7) $\frac{25}{8}$
8) $\frac{1}{8}$
9) 8
10) $\frac{5}{8}$
11) $\frac{49}{8}$
12) $-\frac{1}{2}(x^2 - \frac{3}{2})$
13) 1
14) -2
15) 2
16) 6
17) -1
18) $-2x$

WWW.MathNotion.Com

AFOQT Subject Test – Mathematics

19) 3   21) 2   23) -7
20) -9 22) $\sqrt{13}$ 24) -18

Quadratic Equations

1) $x^2 + 2x - 24$
2) $x^2 + 12x + 35$
3) $x^2 + 2x - 48$
4) $x^2 - 7x - 18$
5) $x^2 - 15x + 56$
6) $3x^2 - 7x - 6$
7) $4x^2 + 5x - 6$
8) $4x^2 - x - 5$
9) $7x^2 - 41x - 6$
10) $15x^2 - 12x - 3$
11) $(x - 4)(x + 2)$
12) $(x + 5)(x + 3)$
13) $(x - 6)(x + 4)$
14) $(x - 3)(x - 7)$
15) $(x + 3)(x + 7)$
16) $(4x + 5)(x + 1)$
17) $(5x - 2)(x + 3)$
18) $(5x - 3)(x + 4)$
19) $(2x + 5)(x + 1)$
20) $3(x - 2)(3x - 1)$
21) $x = -6, x = 3$
22) $x = -1, x = -8$
23) $x = -2, x = -5$
24) $x = 1, x = -2$
25) $x = 3, x = -4$
26) $x = -3, x = -8$
27) $x = -4, x = -\frac{1}{2}$
28) $x = 4, x = -7$
29) $x = 3, x = -4$
30) $x = -3, x = 2$

Solving quadratic equations

1) $\{-9, 1\}$
2) $\{-6, -7\}$
3) $\{8, -3\}$
4) $\{6, 4\}$
5) $\{-2, -12\}$
6) $\{-\frac{4}{5}, -7\}$
7) $\{-\frac{5}{4}, -\frac{1}{6}\}$
8) $\{-\frac{7}{2}, -8\}$
9) $\{-6, -5\}$
10) $\{-\frac{1}{6}, -8\}$
11) $\{8, 0\}$
12) $\{4, -4\}$
13) $\{2, 1\}$
14) $\{-4, -1\}$
15) $\{1, -9\}$
16) $\{2, -12\}$
17) $\{2, -8\}$
18) $\{-3, -6\}$
19) $\{-4, -9\}$
20) $\{5, -3\}$
21) $\{-5, -3\}$
22) $\{1, 3\}$
23) $\{\frac{6}{5}, \frac{3}{2}\}$
24) $\{\frac{6}{7}, 0\}$
25) $\{-\frac{7}{2}, 2\}$
26) $\{\frac{3}{5}, 2\}$
27) $\{-\frac{4}{3}, -4\}$
28) $\{-8, -7\}$
29) $\{4, -1\}$
30) $\{3, -6\}$
31) $\{3, 8\}$
32) $\{\frac{15}{17}, 0\}$

Quadratic formula and the discriminant

1) 576
2) 68
3) 8
4) -11
5) 28
6) 116
7) -11
8) 52
9) 40
10) 69
11) 64
12) 169
13) 336
14) 33
15) 316
16) 132
17) 104
18) 16
19) 1
20) 33
21) 0, one real solution
22) 0, one real solution
23) -783, no solution

WWW.MathNotion.Com

AFOQT Subject Test – Mathematics

24) 0, *one real solution*   26) 0, *one real solution*   28) 0, *one real solution*

25) −127, *no solution*   27) −703, *no solution*

Graphing quadratic functions

1) $(-3, 2), x = -3$

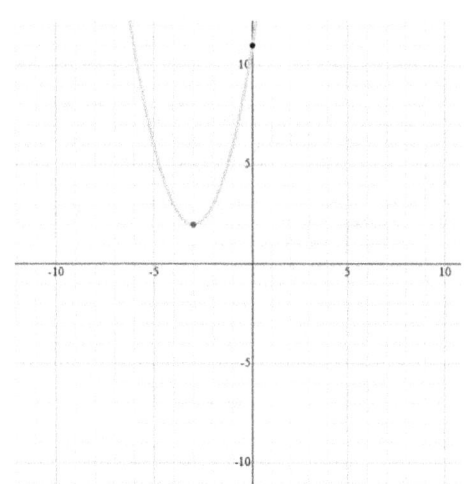

2) $(3, -2), x = 3$

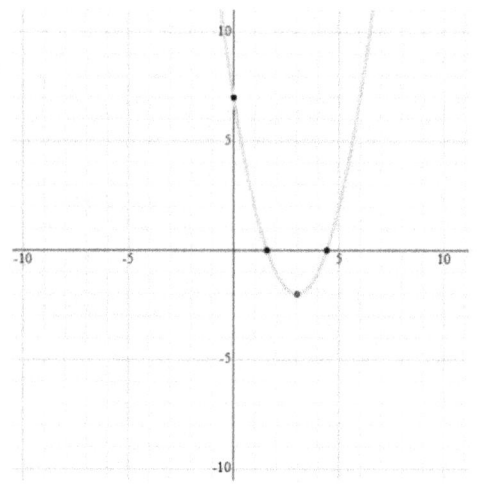

3) $(4, 6), x = 4$

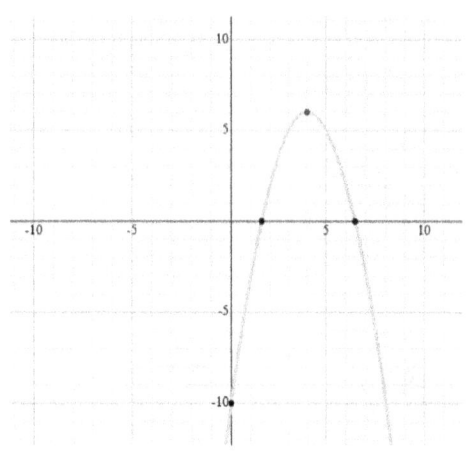

4) $(-1, 12), x = -1$

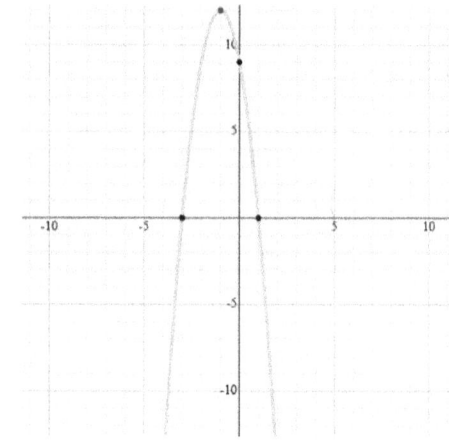

AFOQT Subject Test – Mathematics

Chapter 10 :
Geometry and Solid Figures

Topics that you'll practice in this chapter:

- ✓ Angles
- ✓ Pythagorean Relationship
- ✓ Triangles
- ✓ Polygons
- ✓ Trapezoids
- ✓ Circles
- ✓ Cubes
- ✓ Rectangular Prism
- ✓ Cylinder
- ✓ Pyramids and Cone

Mathematics is, as it were, a sensuous logic, and relates to philosophy as do the arts, music, and plastic art to poetry. — *K. Shegel*

Angles

✏ **What is the value of *x* in the following figures?**

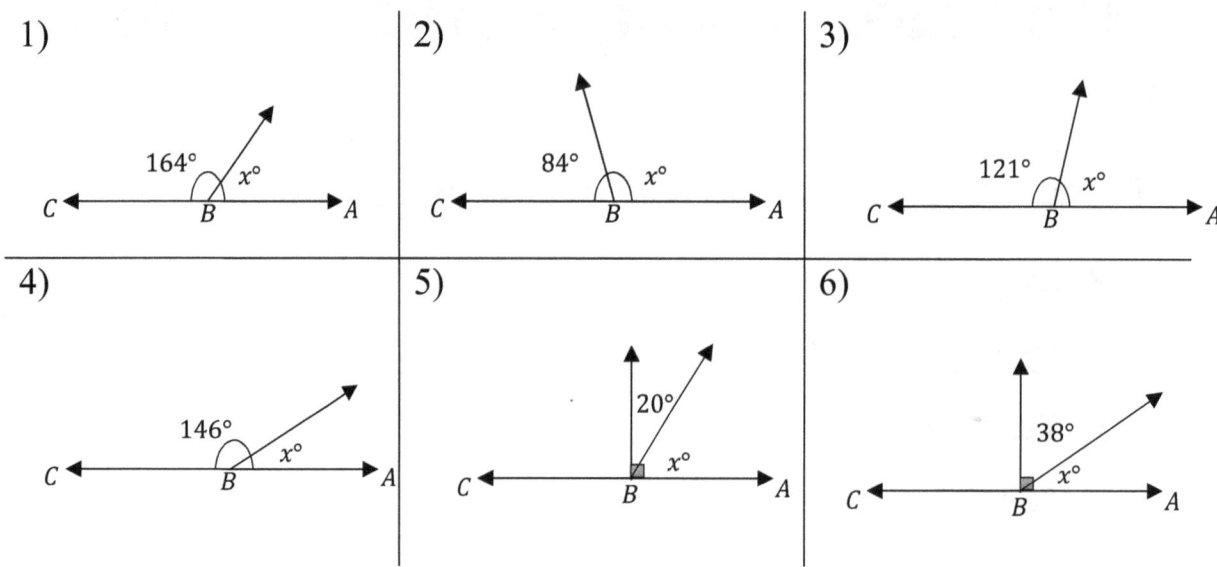

✏ **Calculate.**

7) Two supplement angles have equal measures. What is the measure of each angle? _____

8) The measure of an angle is seven fifth the measure of its supplement. What is the measure of the angle? _____

9) Two angles are complementary and the measure of one angle is 24 less than the other. What is the measure of the smaller angle? _____

10) Two angles are complementary. The measure of one angle is one fifth the measure of the other. What is the measure of the bigger angle? _____

11) Two supplementary angles are given. The measure of one angle is 40° less than the measure of the other. What does the smaller angle measure? _____

AFOQT Subject Test – Mathematics

Pythagorean Relationship

✎ Do the following lengths form a right triangle?

1)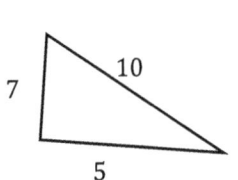

2)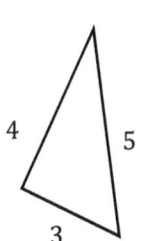

3)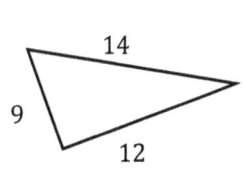

4)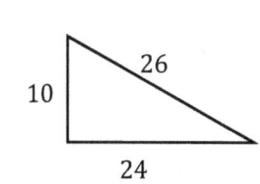

5)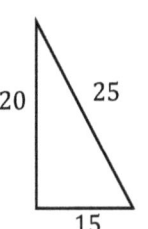

6)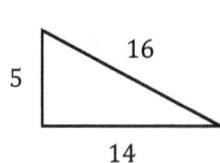

7)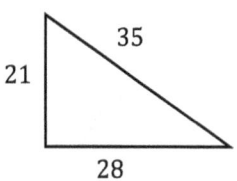

8)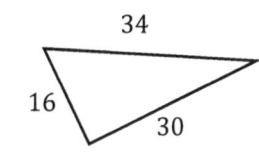

✎ Find the missing side?

9)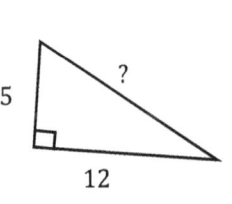

10)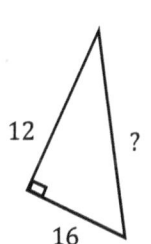

11)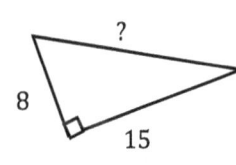

12)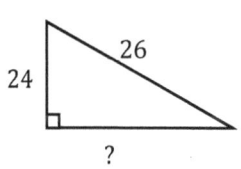

13)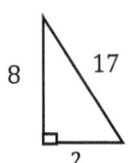

14)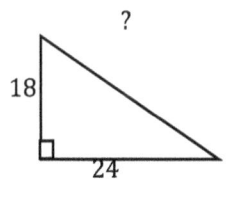

15)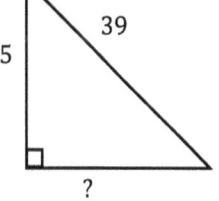

16)

WWW.MathNotion.Com

AFOQT Subject Test – Mathematics

Triangles

✎ Find the measure of the unknown angle in each triangle.

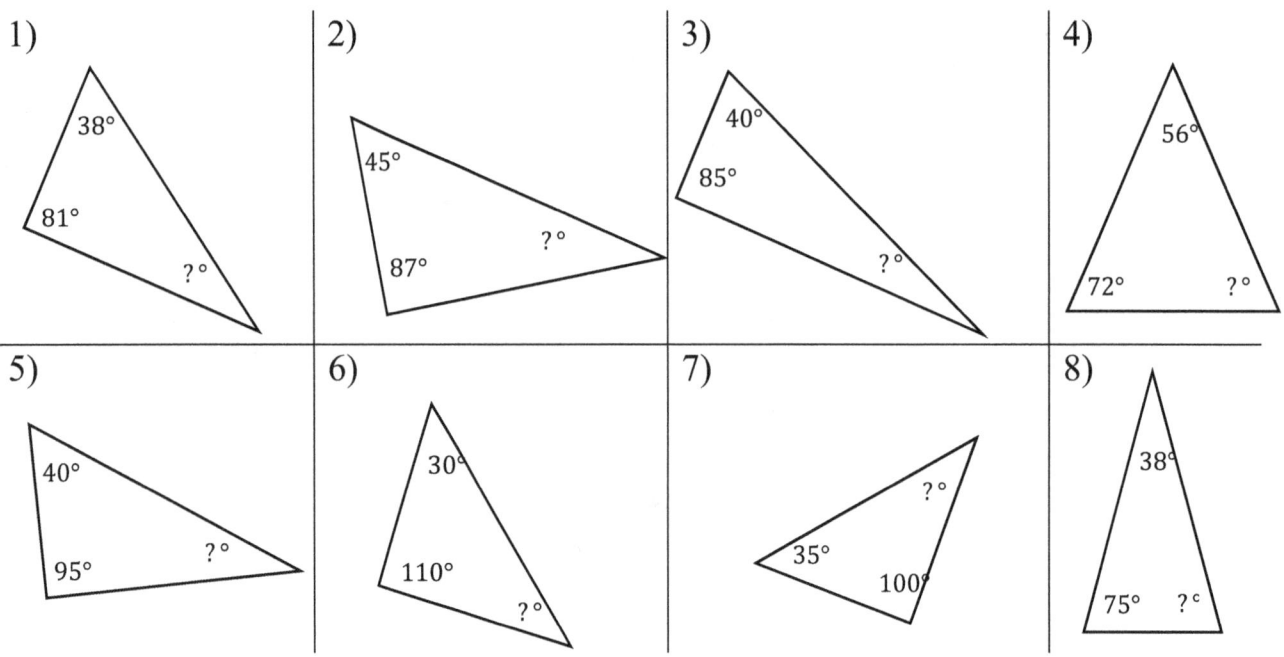

✎ Find area of each triangle.

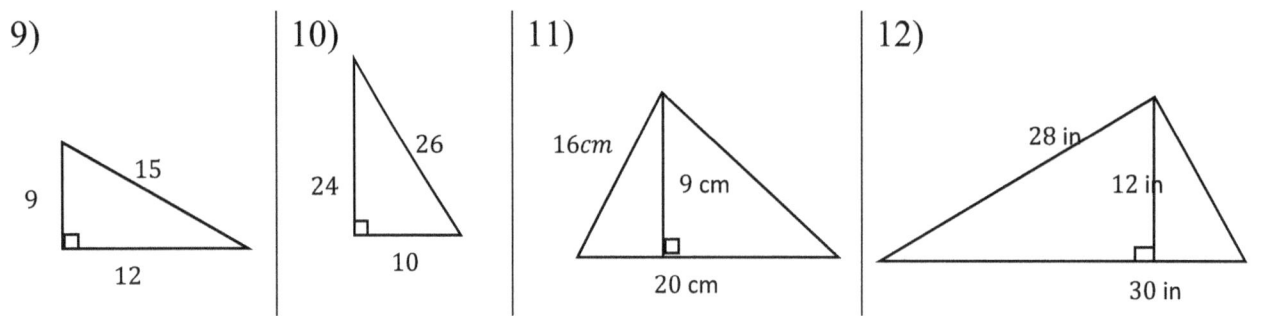

WWW.MathNotion.Com

AFOQT Subject Test – Mathematics

Polygons

🖊 **Find the perimeter of each shape.**

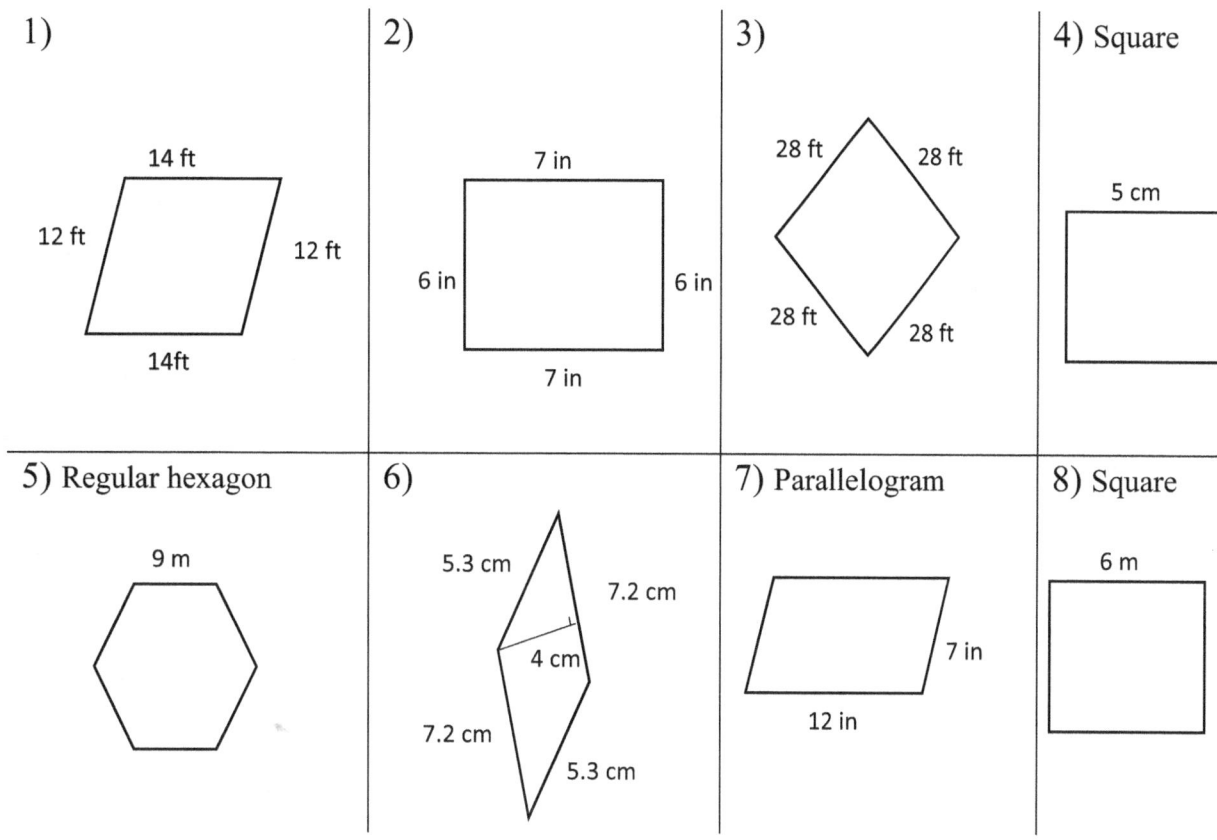

1) 14 ft, 12 ft, 12 ft, 14 ft

2) 7 in, 6 in, 6 in, 7 in

3) 28 ft, 28 ft, 28 ft, 28 ft

4) Square 5 cm

5) Regular hexagon 9 m

6) 5.3 cm, 7.2 cm, 4 cm, 7.2 cm, 5.3 cm

7) Parallelogram 7 in, 12 in

8) Square 6 m

🖊 **Find the area of each shape.**

9) Parallelogram

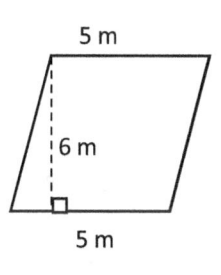

5 m, 6 m, 5 m

10) Rectangle

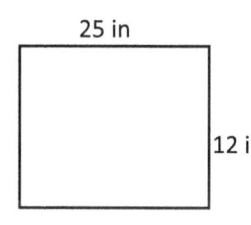

25 in, 12 in

11) Rectangle

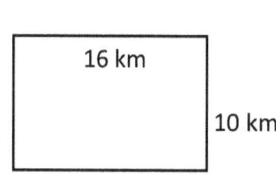

16 km, 10 km

12) Square

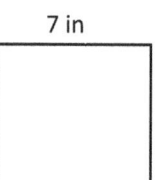

7 in

WWW.MathNotion.Com

AFOQT Subject Test – Mathematics

Trapezoids

✎ Find the area of each trapezoid.

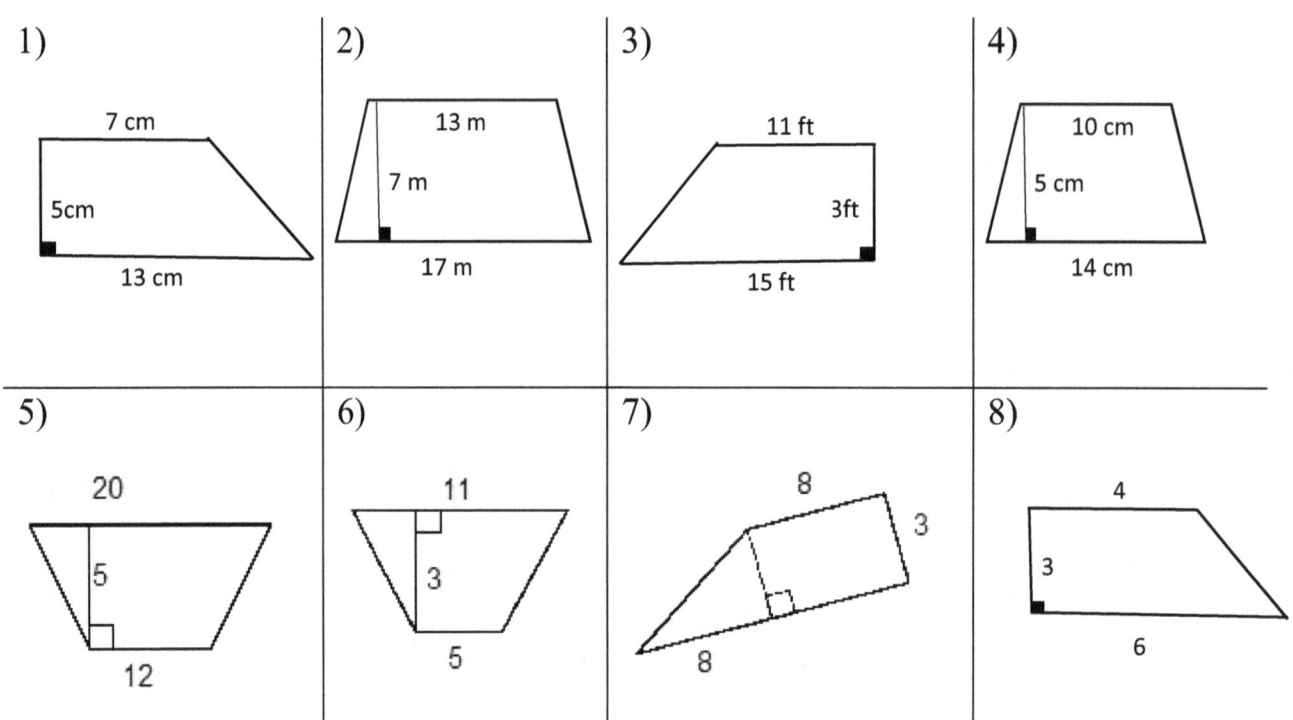

✎ Calculate.

1) A trapezoid has an area of 45 cm² and its height is 5 cm and one base is 5 cm. What is the other base length? _____

2) If a trapezoid has an area of 99 ft² and the lengths of the bases are 8 ft and 10 ft, find the height? _____

3) If a trapezoid has an area of 126 m² and its height is 14 m and one base is 6 m, find the other base length? _____

4) The area of a trapezoid is 440 ft² and its height is 22 ft. If one base of the trapezoid is 15 ft, what is the other base length? _____

AFOQT Subject Test – Mathematics

Circles

✍ **Find the area of each circle.** ($\pi = 3.14$)

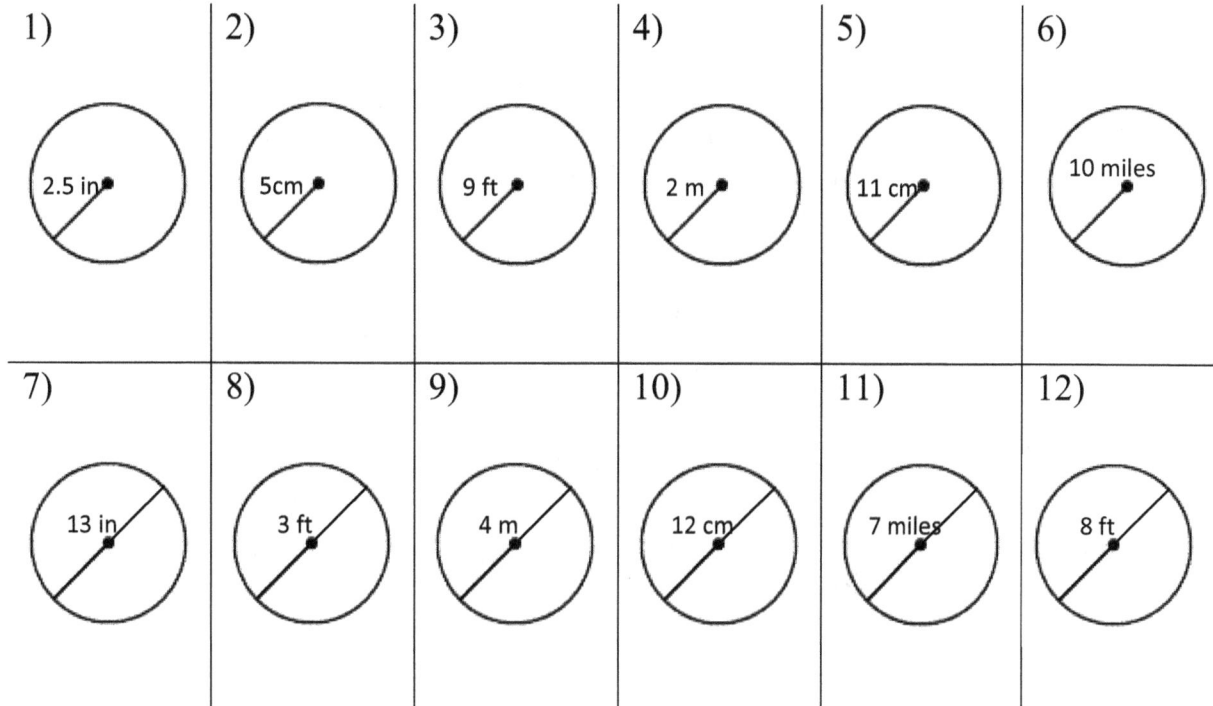

✍ **Complete the table below.** ($\pi = 3.14$)

Circle No.	Radius	Diameter	Circumference	Area
1	1 in	2 in	6.28 in	3.14 in^2
2		10 m		
3				28.26 ft^2
4			47.1 mi	
5		11 km		
6	7 cm			
7		12 ft		
8				314 m^2
9			56.52 in	
10	4.5 ft			

WWW.MathNotion.Com

AFOQT Subject Test – Mathematics

Cubes

✎ Find the volume of each cube.

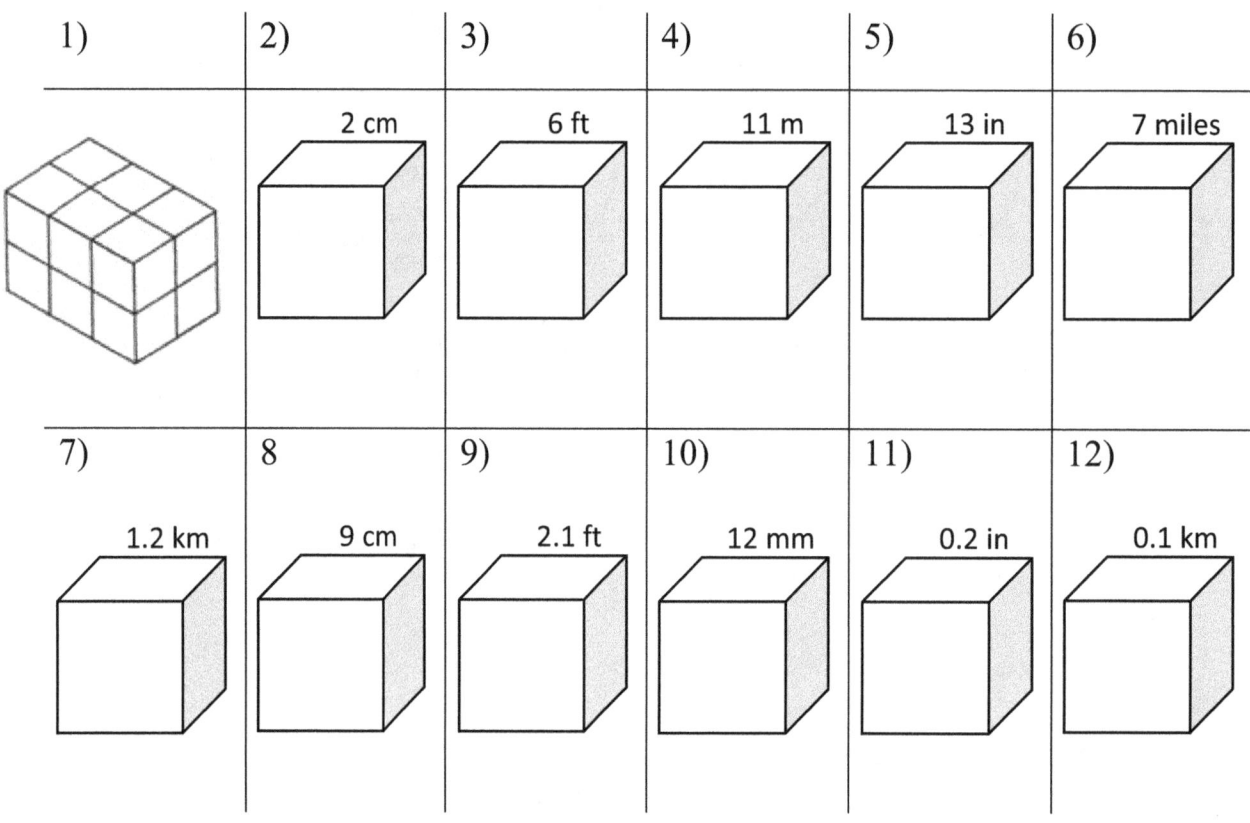

✎ Find the surface area of each cube.

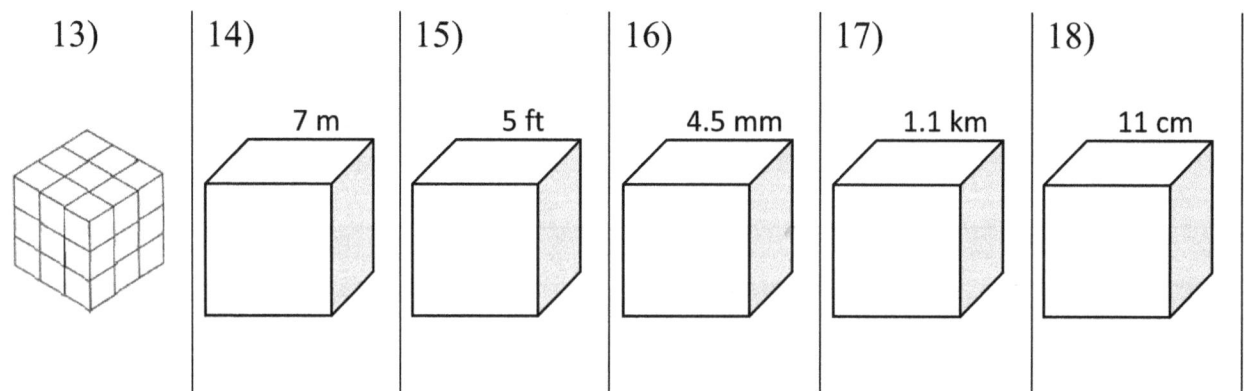

AFOQT Subject Test – Mathematics

Rectangular Prism

✎ Find the volume of each Rectangular Prism.

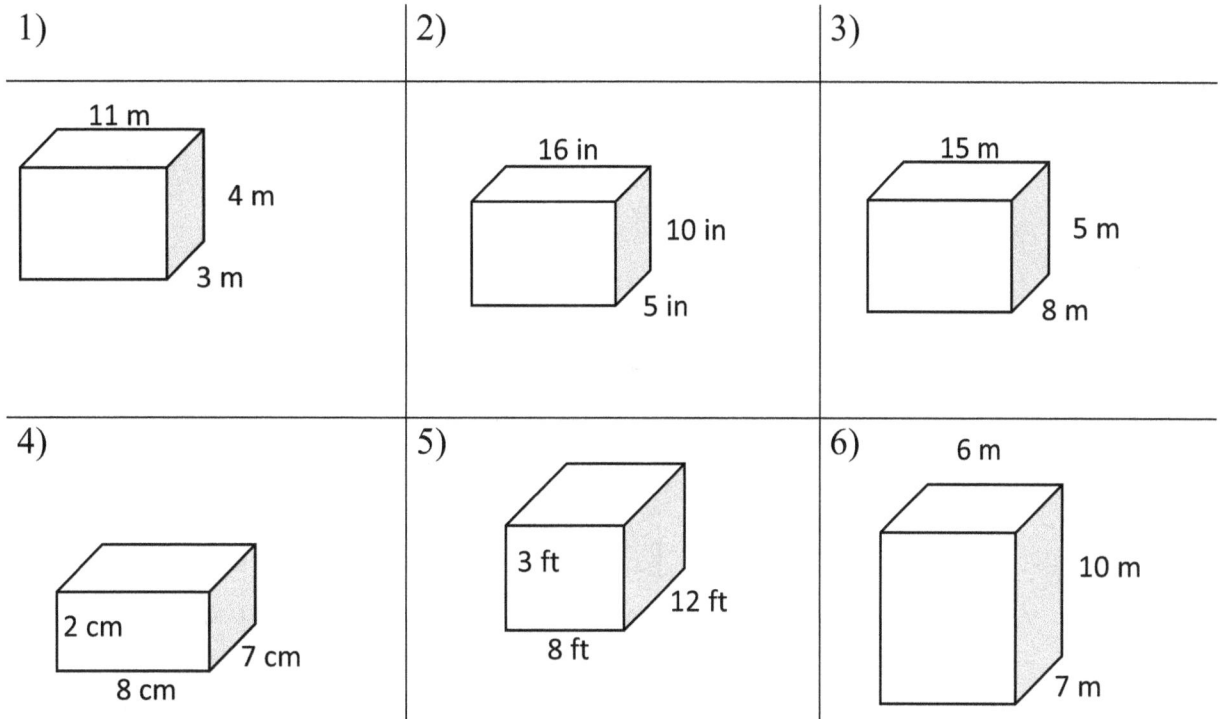

✎ Find the surface area of each Rectangular Prism.

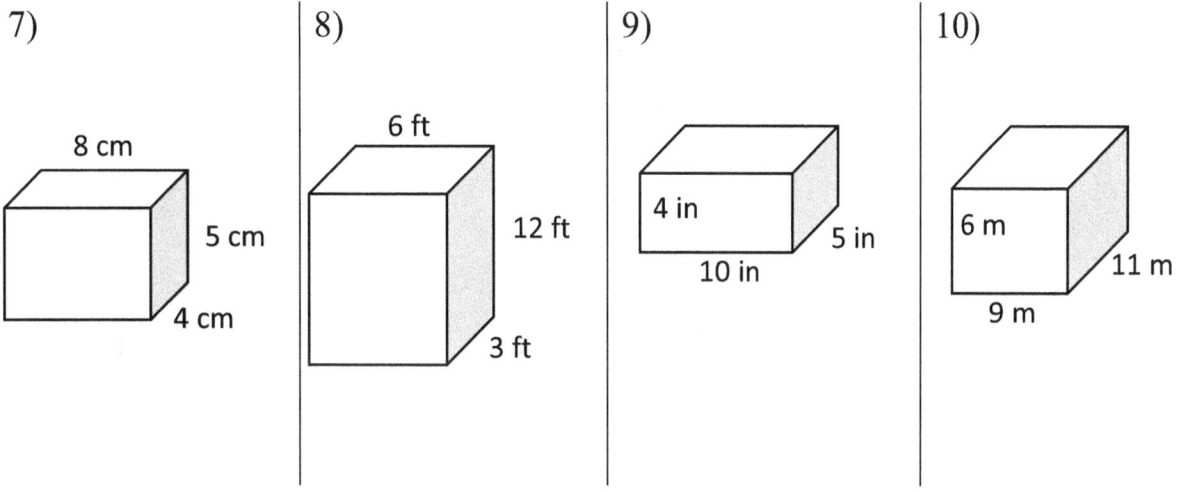

AFOQT Subject Test – Mathematics

Cylinder

✏️ **Find the volume of each Cylinder. Round your answer to the nearest tenth.** ($\pi = 3.14$)

1)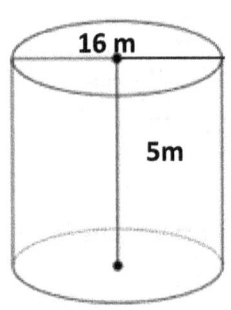

2)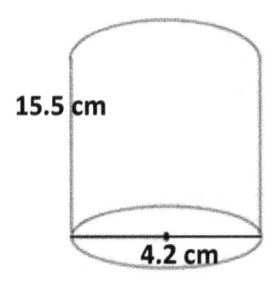

3)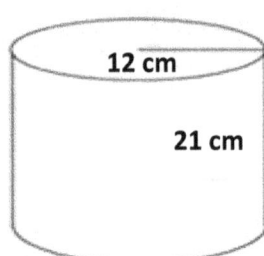

4)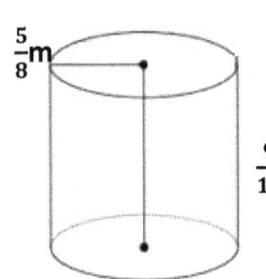

5)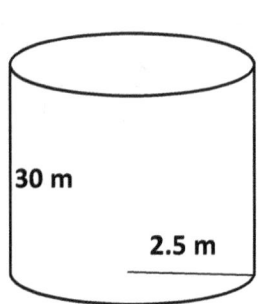

6)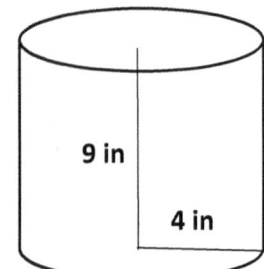

✏️ **Find the surface area of each Cylinder.** ($\pi = 3.14$)

7)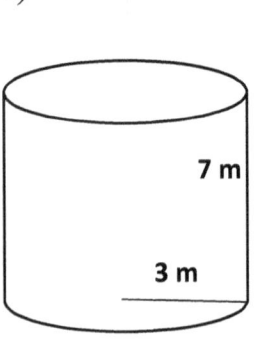

8)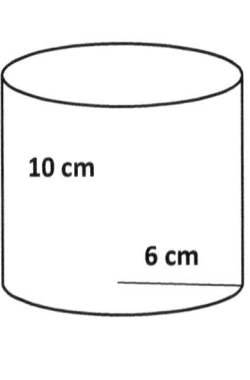

9)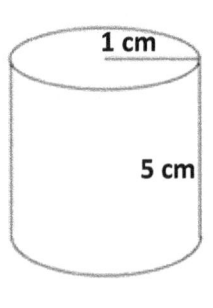

10)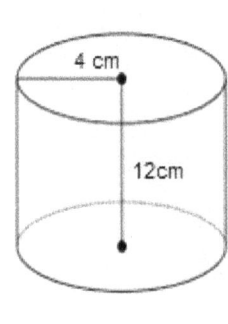

WWW.MathNotion.Com

AFOQT Subject Test – Mathematics

Pyramids and Cone

✎ Find the volume of each Pyramid and Cone. ($\pi = 3.14$)

1)
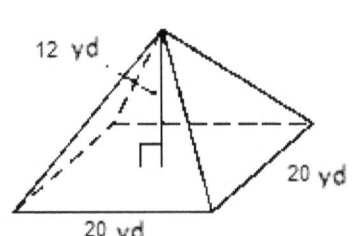

2)
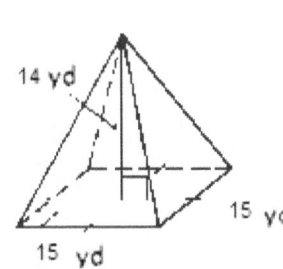

3)
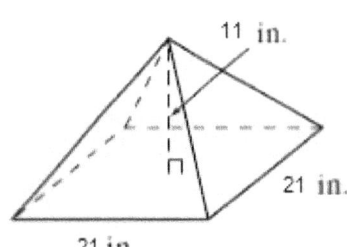

4)
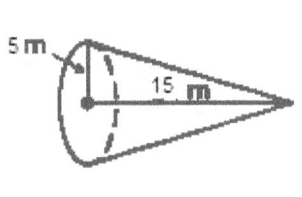

5)
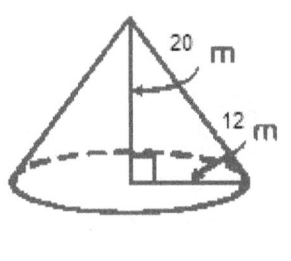

6)
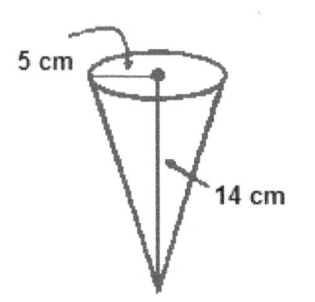

✎ Find the surface area of each Pyramid and Cone. ($\pi = 3.14$)

7)

8)

9)

10)

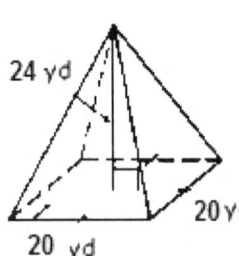
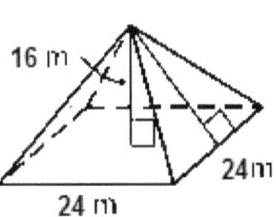
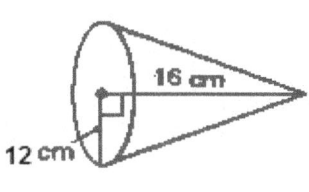

WWW.MathNotion.Com

AFOQT Subject Test – Mathematics

Answers of Worksheets

Angles
1) 16°
2) 96°
3) 59°
4) 34°
5) 70°
6) 52°
7) 90°
8) 75°
9) 33°
10) 75°
11) 70°

Pythagorean Relationship
1) No
2) Yes
3) No
4) Yes
5) Yes
6) No
7) Yes
8) Yes
9) 13
10) 20
11) 17
12) 10
13) 15
14) 30
15) 36
16) 12

Triangles
1) 60°
2) 48°
3) 55°
4) 52°
5) 45°
6) 40°
7) 45°
8) 67°
9) 54 square unites
10) 120 square unites
11) 90 square unites
12) 180 square unites

Polygons
1) 52 ft
2) 26 in
3) 112 ft
4) 20 cm
5) 54 m
6) 25 cm
7) 38 in
8) 24 m
9) 30 m^2
10) 300 in^2
11) 160 km^2
12) 49 in^2

Trapezoids
1) 50 cm^2
2) 105 m^2
3) 39 ft^2
4) 60 cm^2
5) 80
6) 24
7) 36
8) 15

Calculate
1) 13 cm
2) 11 ft
3) 12 m
4) 25 ft

Circles
1) 19.63 in^2
2) 78.5 cm^2
3) 254.34 ft^2
4) 12.56 m^2
5) 379.94 cm^2
6) 314 $miles^2$
7) 132.67 in^2
8) 7.07 ft^2
9) 12.56 m^2
10) 113.04 cm^2
11) 38.47 $miles^2$
12) 50.24 ft^2

AFOQT Subject Test – Mathematics

Circle No.	Radius	Diameter	Circumference	Area
1	1 in	2 in	6.28 in	3.14 in^2
2	5 m	10 m	31.4 m	78.5 m^2
3	3 ft	6 ft	18.84 ft	28.26 ft^2
4	7.5 miles	15 mi	47.1 mi	176.63 mi^2
5	5.5 km	11 km	34.54 km	94.99 km^2
6	7 cm	14 cm	43.96 cm	153.86 cm^2
7	6 ft	12 ft	37.68 feet	113.04 ft^2
8	10 m	20 m	62.8 m	314 m^2
9	9 in	18 in	56.52 in	254.34 in^2
10	4.5 ft	9 ft	28.26 ft	63.585 ft^2

Cubes

1) 12
2) 8 cm^3
3) 216 ft^3
4) 1,331 m^3
5) 2,197 in^3
6) 343 $miles^3$
7) 1.728 km^3
8) 729 cm^3
9) 9.261 ft^3
10) 1,728 mm^3
11) 0.008 in^3
12) 0.001 km^3
13) 27
14) 294 m^2
15) 150 ft^2
16) 121.5 mm^2
17) 7.26 km^2
18) 726 cm^2

Rectangular Prism

1) 132 m^3
2) 800 in^3
3) 600 m^3
4) 112 cm^3
5) 288 ft^3
6) 420 m^3
7) 184 cm^2
8) 252 ft^2
9) 220 in^2
10) 438 m^2

Cylinder

1) 1,004.8 m^3
2) 214.6 cm^3
3) 9,495.4 cm^3
4) 1.1 m^3
5) 588.8 m^3
6) 452.2 in^3
7) 188.4 m^2
8) 602.9 cm^2
9) 37.7 cm^2
10) 401.9 m^2

Pyramids and Cone

1) 1,600 yd^3
2) 1,050 yd^3
3) 1,617 in^3
4) 392.5 m^3
5) 3,014.4 m^3
6) 366.33 cm^3
7) 1,440 yd^2
8) 1,536 m^2
9) 678.24 in^2
10) 1,205.76 cm^2

AFOQT Subject Test – Mathematics

AFOQT Subject Test – Mathematics

Chapter 11 :
Statistics and Probability

Topics that you'll practice in this chapter:

- ✓ Mean and Median
- ✓ Mode and Range
- ✓ Histograms
- ✓ Stem–and–Leaf Plot
- ✓ Pie Graph
- ✓ Probability Problems

Mathematics is no more computation than typing is literature.

– John Allen Paulos

AFOQT Subject Test – Mathematics

Mean and Median

✒ **Find Mean and Median of the Given Data.**

1) 8, 7, 14, 4, 8

2) 14, 8, 25, 19, 16, 33, 11

3) 23, 18, 15, 12, 17

4) 34, 14, 10, 15, 6, 11

5) 10, 19, 6, 8, 32, 20, 17

6) 17, 26, 39, 69, 20, 6

7) 40, 38, 18, 11, 9, 2, 7, 32, 41

8) 24, 21, 31, 12, 33, 32, 22

9) 16, 14, 20, 41, 15, 20, 38, 4

10) 20, 20, 30, 18, 6, 28, 12, 46

11) 12, 7, 10, 11, 16, 22

12) 10, 29, 27, 12, 2, 15, 10, 3

✒ **Calculate.**

13) In a javelin throw competition, five athletics score 56, 34, 62, 23 and 19 meters. What are their Mean and Median? _____

14) Eva went to shop and bought 8 apples, 14 peaches, 6 bananas, 4 pineapples and 12 melons. What are the Mean and Median of her purchase? _____

15) Bob has 17 black pen, 19 red pen, 14 green pens, 20 blue pens and 5 boxes of yellow pens. If the Mean and Median are 19 respectively, what is the number of yellow pens in each box? _____

AFOQT Subject Test – Mathematics

Mode and Range

✎ **Find Mode and Rage of the Given Data.**

1) 4, 3, 7, 3, 3, 4
   Mode: _____   Range: _____

2) 18, 18, 24, 26, 18, 8, 14, 22
   Mode: _____   Range: _____

3) 8, 8, 8, 16, 19, 22, 20, 9, 13
   Mode: _____   Range: _____

4) 24, 24, 14, 28, 20, 18, 20, 24
   Mode: _____   Range: _____

5) 6, 21, 27, 24, 27, 27
   Mode: _____   Range: _____

6) 21, 8, 8, 7, 8, 12, 10, 22, 18, 13
   Mode: _____   Range: _____

7) 7, 4, 4, 6, 13, 13, 13, 0, 2, 2
   Mode: _____   Range: _____

8) 5, 8, 5, 14, 12, 14, 3, 5, 18
   Mode: _____   Range: _____

9) 7, 7, 7, 12, 7, 3, 8, 16, 3, 17
   Mode: _____   Range: _____

10) 15, 15, 19, 16, 4, 16, 10, 15
    Mode: _____   Range: _____

11) 6, 6, 5, 6, 42, 13, 19, 2
    Mode: _____   Range: _____

12) 8, 8, 9, 8, 9, 4, 34, 22
    Mode: _____   Range: _____

✎ **Calculate.**

13) A stationery sold 12 pencils, 56 red pens, 24 blue pens, 20 notebooks, 12 erasers, 21 rulers and 11 color pencils. What are the Mode and Range for the stationery sells?

    Mode: _____   Range: _____

14) In an English test, eight students score 10, 15, 15, 18 18, 16, 15 and 15. What are their Mode and Range? _____

15) What is the range of the first 6 even numbers greater than 8?
    _____

WWW.MathNotion.Com

AFOQT Subject Test – Mathematics

Times Series

✎ Use the following Graph to complete the table.

Day	Distance (km)
1	
2	

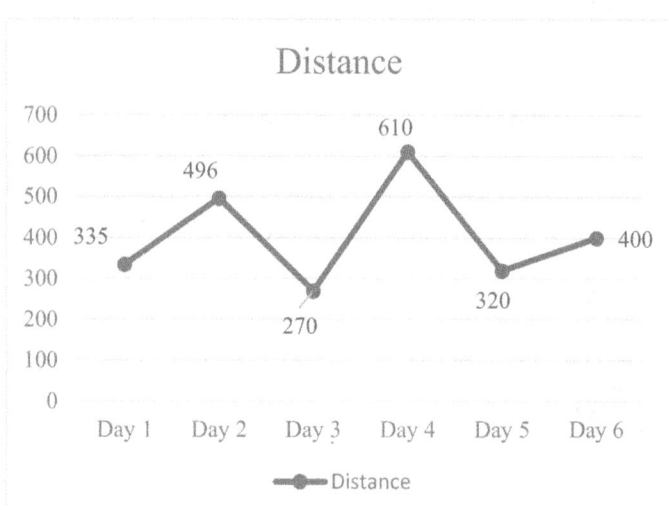

The following table shows the number of births in the US from 2007 to 2012 (in millions).

Year	Number of births (in millions)
2007	4.15
2008	3.70
2009	3.45
2010	3.20
2011	1.75
2012	2.98

Draw a Time Series for the table.

AFOQT Subject Test – Mathematics

Stem-and-Leaf Plot

✎ **Make stem ad leaf plots for the given data.**

1) 24, 26, 29, 20, 53, 27, 51, 55, 36, 21, 37, 30

   Stem | Leaf plot

2) 11, 59, 66, 14, 18, 19, 59, 65, 69, 61, 68, 65

   Stem | Leaf plot

3) 121, 55, 66, 54, 112, 128, 63, 125, 59, 123, 68, 119

   Stem | Leaf plot

4) 51, 32, 100, 56, 84, 36, 107, 56, 85, 39, 56, 106, 89

   Stem | Leaf plot

5) 33, 89, 19, 87, 81, 16, 11, 30, 86, 35, 17, 35, 13

   Stem | Leaf plot

6) 60, 92, 22, 25, 67, 93, 95, 62, 21, 64, 98, 29

   Stem | Leaf plot

WWW.MathNotion.Com

AFOQT Subject Test – Mathematics

Pie Graph

The circle graph below shows all Robert's expenses for last month. Robert spent $140 on his hobbies last month.

Answer following questions based on the Pie graph.

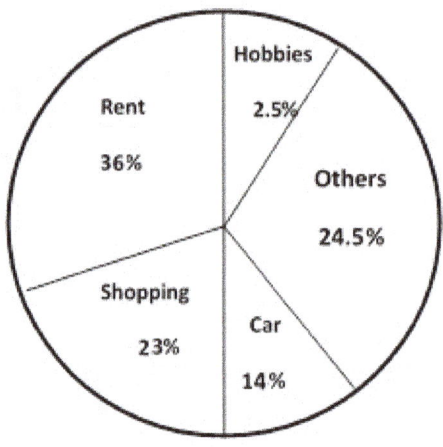

1) How much was Robert's total expenses last month? _____

2) How much did Robert spend on his car last month? _____

3) How much did Robert spend for shopping last month? _____

4) How much did Robert spend on his rent last month? _____

5) What fraction is Robert's expenses for his rent and car out of his total expenses last month? _____

AFOQT Subject Test – Mathematics

Probability Problems

✎ **Calculate.**

1) A number is chosen at random from 1 to 10. Find the probability of selecting number 6 or smaller numbers. _____

2) Bag A contains 18 red marbles and 6 green marbles. Bag B contains 16 black marbles and 8 orange marbles. What is the probability of selecting a green marble at random from bag A? What is the probability of selecting a black marble at random from Bag B? _____

3) A number is chosen at random from 1 to 20. What is the probability of selecting multiples of 4? _____

4) A card is chosen from a well-shuffled deck of 52 cards. What is the probability that the card will be a queen? _____

5) A number is chosen at random from 1 to 15. What is the probability of selecting a multiple of 3 or 5? _____

A spinner numbered 1–8, is spun once. What is the probability of spinning …?

6) an Odd number? _____   7) a multiple of 2? _____

8) a multiple of 5? _____   9) number 10? _____

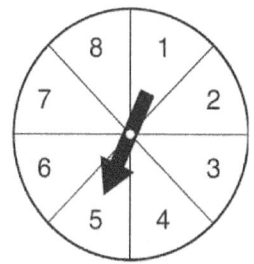

WWW.MathNotion.Com

AFOQT Subject Test – Mathematics

Answers of Worksheets

Mean and Median

1) Mean: 8.2, Median: 8
2) Mean: 18, Median: 16
3) Mean: 17, Median: 17
4) Mean: 15, Median: 12.5
5) Mean: 16, Median: 17
6) Mean: 29.5, Median: 23
7) Mean: 22, Median: 18
8) Mean: 25, Median: 24
9) Mean: 21, Median: 18
10) Mean: 22.5, Median: 20
11) Mean: 13, Median: 11.5
12) Mean: 13.5, Median: 11
13) Mean: 38.8, Median: 34
14) Mean: 8.8, Median: 8
15) 5

Mode and Range

1) Mode: 3, Range: 4
2) Mode: 18, Range: 18
3) Mode: 8, Range: 14
4) Mode: 24, Range: 14
5) Mode: 27, Range: 21
6) Mode: 8, Range: 15
7) Mode: 13, Range: 13
8) Mode: 5, Range: 15
9) Mode: 7, Range: 14
10) Mode: 15, Range: 15
11) Mode: 6, Range: 40
12) Mode: 8, Range: 30
13) Mode: 12, Range: 45
14) Mode: 15, Range: 8
15) 10

Time series

Day	Distance (km)
1	335
2	496
3	270
4	610
5	320
6	400

Stem–And–Leaf Plot

1)

Stem	leaf
2	0 1 4 6 7 9
3	0 6 7
5	1 3 5

2)

Stem	leaf
1	1 4 8 9
5	9 9
6	1 5 5 6 8 9

3)

Stem	leaf
5	4 5 9
6	3 6 8
11	2 9
12	1 3 5 8

AFOQT Subject Test – Mathematics

4)

Stem	leaf
3	2 6 9
5	1 6 6 6
8	4 5 9
10	0 6 7

5)

Stem	leaf
1	1 3 6 7 9
3	0 3 5 5
8	1 6 7 9

6)

Stem	leaf
2	2 1 5 9
6	0 2 4 7
9	2 3 5 8

Pie Graph

1) $5,600   3) $1,288   5) $\frac{1}{2}$

2) $784   4) $2,016

Probability Problems

1) $\frac{3}{5}$ 4) $\frac{1}{13}$ 7) $\frac{1}{2}$

2) $\frac{1}{4}, \frac{2}{3}$ 5) $\frac{7}{15}$ 8) $\frac{1}{8}$

3) $\frac{1}{4}$ 6) $\frac{1}{2}$ 9) 0

AFOQT Subject Test – Mathematics

AFOQT Subject Test – Mathematics

Chapter 12 : AFOQT Math Practice Tests

The Air Force Officer Qualifying Test (AFOQT) is a standardized test to assess skills and personality traits that have proven to be predictive of success in officer commissioning programs such as the training program.

The AFOQT is used to select applicants for officer commissioning programs, such as Officer Training School (OTS) or Air Force Reserve Officer Training Corps (Air Force ROTC) and pilot and navigator training.

The AFOQT is a multiple-aptitude battery that measures developed abilities and helps predict future academic and occupational success in the military. The AFOQT is a multiple-choice test which consists of 12 subtests and two of them are Arithmetic Reasoning and Mathematics Knowledge.

In this section, there are 2 complete Arithmetic Reasoning and Mathematics Knowledge AFOQT Tests. Take these tests to see what score you'll be able to receive on a real AFOQT test.

Good Luck

AFOQT Subject Test – Mathematics

AFOQT Subject Test – Mathematics

AFOQT Practice Test 1

Arithmetic Reasoning

- 25 questions
- Total time for this section: 29 Minutes
- Calculators are not allowed for this test.

Administered *Month Year*

AFOQT Subject Test – Mathematics

1) What is the product of the square root of 144 and the square root of 25?

   A. 3,600
   B. 60
   C. 720
   D. 120

2) A bread recipe calls for $5\frac{1}{14}$ cups of flour. If you only have $4\frac{6}{7}$ cups of flour, how much more flour is needed?

   A. $1\frac{1}{14}$
   B. $\frac{3}{14}$
   C. $1\frac{3}{14}$
   D. $\frac{1}{14}$

3) There are 245 rooms that need to be painted and only 14 painters available. If there are still 35 rooms unpainted by the end of the day, what is the average number of rooms that each painter has painted?

   A. 5
   B. 15
   C. 10
   D. 25

4) Convert 0.034 to a percent.

   A. 0.34%
   B. 0.43%
   C. 3.40%
   D. 34%

5) If $5x - 24x + 19x = -32$, then what is the value of x?

   A. Any Number
   B. Positive Infinity
   C. Does not exit
   D. Negative Infinity

AFOQT Subject Test – Mathematics

6) Mia is 8 years older than her sister Elise, and Elise is 6 years younger than her brother Mason. If the sum of their ages is 110, how old is Elise?

   A. 32

   B. 33

   C. 40

   D. 38

7) John is driving to visit his mother, who lives 175 miles away. How long will the drive be, round-trip, if John drives at an average speed of 25 mph?

   A. 70 Minutes

   B. 140 Minutes

   C. 260 Minutes

   D. 420 Minutes

8) While at work, Emma checks her email once every 30 minutes. In 9-hour, how many times does she check her email?

   A. 6 Times

   B. 12 Times

   C. 24 Times

   D. 18 Times

9) Julie gives 8 pieces of candy to each of her friends. If Julie gives all her candy away, which amount of candy could have been the amount she distributed?

   A. 430

   B. 584

   C. 326

   D. 644

10) If a rectangle is 52 feet by 30 feet, what is its area?

    A. 2,600

    B. 1,560

    C. 1,532

    D. 780

AFOQT Subject Test – Mathematics

11) You are asked to chart the temperature during a 6-hour period to give the average. These are your results:

   6 am: 10 degrees   1 pm: 27 degrees

   8 am: 13 degrees   3 pm: 32 degrees

   10 am: 21 degrees   5 pm: 29 degrees

   What is the average temperature?

   A. 19   C. 22

   B. 18   D. 24

12) Each year, a cybercafé charges its customers a base rate of $35, with an additional $0.40 per visit for the first 60 visits, and $0.30 for every visit after that. How much does the cybercafé charge a customer for a year in which 100 visits are made?

   A. $47   C. $59

   B. $83   D. $71

13) If a vehicle is driven 31 miles on Monday, 42 miles on Tuesday, and 14 miles on Wednesday, what is the average number of miles driven each day?

   A. 29 Miles   C. 27 Miles

   B. 24 Miles   D. 32 Miles

14) What is the prime factorization of 800?

   A. $2 \times 2 \times 3 \times 3$ C. $2 \times 3 \times 17$

   B. $2 \times 2 \times 2 \times 2 \times 2 \times 5 \times 5$ D. $2 \times 3 \times 5 \times 7$

AFOQT Subject Test – Mathematics

15) Three co-workers contributed $14.36, $12.41, and $19.53 respectively to purchase a retirement gift for their boss. What is the maximum amount they can spend on a gift?

   A. 54.21

   B. $46.30

   C. $16.72

   D. $28.36

16) A family owns 15 dozen of magazines. After donating 52 magazines to the public library, how many magazines are still with the family?

   A. 780

   B. 113

   C. 128

   D. 624

17) In the deck of cards, there are 4 spades, 13 hearts, 8 clubs, and 7 diamonds. What is the probability that William will pick out a spade?

   A. $\frac{1}{9}$

   B. $\frac{1}{8}$

   C. $\frac{1}{3}$

   D. $\frac{13}{32}$

18) William is driving a truck that can hold 15 tons maximum. He has a shipment of food weighing 75,000 pounds. How many trips will he need to make to deliver all the food?

   A. 3 Trip

   B. 2.5 Trips

   C. 1.5 Trips

   D. 3.5 Trips

AFOQT Subject Test – Mathematics

19) A man goes to a casino with $460. He loses $70 on blackjack, then loses another $150 on roulette. How much money does he have left?

   A. $230

   B. $240

   C. $235

   D. $280

20) A woman owns a dog walking business. If 5 workers can walk 15 dogs, how many dogs can 7 workers walk?

   A. 7

   B. 20

   C. 3

   D. 21

21) Jude was hired to teach six identical math courses, which entailed being present in the classroom 54 hours altogether. At $35 per class hour, how much did Aria earn for teaching one course?

   A. $210

   B. $210

   C. $324

   D. $315

22) If one acre of forest contains 112 pine trees, how many pine trees are contained in 25 acres?

   A. 5

   B. 2,800

   C. 8

   D. 137

23) Ava needs $\frac{1}{7}$ of an ounce of salt to make 1 cup of dip for fries. How many cups of dip will she be able to make if she has 35 ounces of salt?

   A. 7

   B. $\frac{1}{7}$

AFOQT Subject Test – Mathematics

C. 245

D. $\frac{1}{245}$

24) Six out of 60 students had to go to summer school. What is the ratio of students who did not have to go to summer school expressed, in its lowest terms?

A. $\frac{9}{10}$

B. $\frac{1}{10}$

C. $1\frac{1}{10}$

D. $1\frac{9}{10}$

25) I've got 64 quarts of milk and my family drinks 2 gallons of milk per week. How many weeks will that last us?

A. 18 Weeks

B. 4 Weeks

C. 4.5 Weeks

D. 8 Weeks

AFOQT Subject Test – Mathematics

AFOQT Subject Test – Mathematics

AFOQT Practice Test 1

Mathematics Knowledge

- 25 questions
- Total time for this section: 22 Minutes
- Calculators are not allowed for this test.

Administered Month Year

AFOQT Subject Test – Mathematics

1) Which of the following is true equal to 6^3?

   A. the square of 6

   B. 6 squared

   C. 6 cubed

   D. 6 to the second power

2) What is the reciprocal of $\frac{x^4}{81}$?

   A. $\frac{3}{x^4} - 4$

   B. $\frac{4}{x^4}$

   C. $\frac{3}{x^4} + 4$

   D. $(\frac{3}{x})^4$

3) If a = 12, what is the value of b in this equation? $b = \frac{a^2}{6} + 5$

   A. 17

   B. 24

   C. 30

   D. 29

4) The sixth root of 729 is:

   A. 6

   B. 12

   C. 3

   D. 9

5) A circle has a radius of 5 inches. What is its approximate area? ($\pi = 3.14$)

   A. 314 square inches

   B. 78.50 square inches

   C. 49.29 square inches

   D. 31.40 square inches

6) In the following diagram what is the value of x?

   A. 63

   B. 27

   C. 153

   D. 27

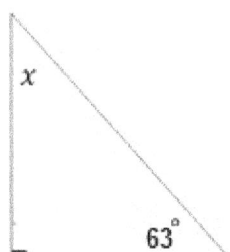

AFOQT Subject Test – Mathematics

7) If $-15a = 165$, then $a = $ ___

   A. -11

   B. 15

   C. 11

   D. 7

8) What is 361,297,580 in scientific notation?

   A. 36.129758

   B. 3.6129758×10^8

   C. 0.036129758×10^8

   D. 0.36129758

9) In the following right triangle, what is the value of x rounded to the nearest hundredth?

   A. 12

   B. 16

   C. 26

   D. 18.20

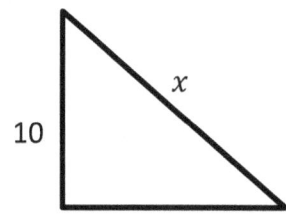

10) Which of the following sets of factors do both 16 and 36 have in common?

    A. {0, 2, 6, 16}

    B. {1, 2, 4}

    C. {0, 2, 4}

    D. {2, 3, 4}

11) Which of the following is a composite number?

    A. 39

    B. 61

    C. 43

    D. 53

12) $(x + 4)(2x + 5) = $?

    A. $13x^2 + 20$

    B. $2x^2 + 20x + 13$

    C. $2x^2 + 13x + 20$

    D. $2x^2 + 20$

AFOQT Subject Test – Mathematics

13) $11(a - 4) = 33$, what is the value of a?

   A. 4

   B. 8

   C. 9

   D. 7

14) If $16^{11} = 4^7 \times 4^{3x}$, what is the value of x?

   A. 1

   B. 6

   C. 2

   D. 5

15) The volume of this box is:

   A. 313 cm³

   B. 330 cm³

   C. 310 cm³

   D. 325 cm³

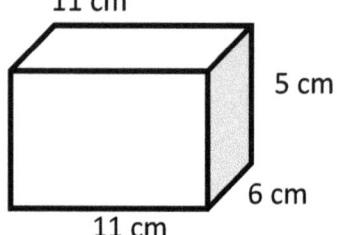

16) Find the slope of the line running through the points (4, 2) and (5, 8).

   A. $-\frac{1}{6}$

   B. -6

   C. 6

   D. $\frac{1}{6}$

17) In the following diagram, the straight line is divided by one angled line at 58°. What is the value of ∠a.

   A. 32°

   B. 22°

   C. 122°

   D. 158°

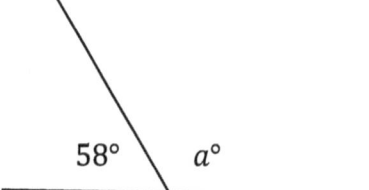

AFOQT Subject Test – Mathematics

18) Factor this expression: $x^2 - 4x - 32$

   A. $x^2(4x + 8)$

   B. $4x(x - 8)$

   C. $(x - 8)(x + 4)$

   D. $(x + 8)(x - 4)$

19) What's the area of the non-shaded part in the following figure?

   A. 116

   B. 76

   C. 96

   D. 20

20) A medium pizza has a diameter of 10 inches. What is its area?

   A. 10π

   B. 100π

   C. 5π

   D. 25π

21) What is the circumference of a circle with center at point A if the distance from point X to Y is 16 feet? ($\pi = 3.14$)

   A. 12.56 Feet

   B. 50.24 Feet

   C. 200.96 Feet

   D. 25.12 Feet

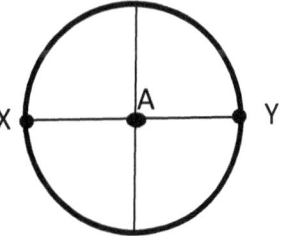

22) Solve for the value of y in the following system of equations.

$$5x - 3y = -4, -3x + y = 4$$

   A. 2

   B. -2

   C. -4

   D. 4

AFOQT Subject Test – Mathematics

23) What is the value of $\sqrt{18} \times \sqrt{72}$?

    A. $18\sqrt{3}$ C. 36

    B. $18\sqrt{2}$ D. $36\sqrt{2}$

24) Which of the following is an acute angle?

    A. 95°   C. 100°

    B. 160°   D. 79°

25) What is the value of $\frac{8!}{5!}$?

    A. 56   C. 40

    B. 336   D. 1,680

AFOQT Subject Test – Mathematics

AFOQT Practice Test 2

Arithmetic Reasoning

- 25 questions
- Total time for this section: 29 Minutes
- Calculators are not allowed for this test.

Administered Month Year

AFOQT Subject Test – Mathematics

1) Camille uses a 25% off coupon when buying a sweater that costs $160. If she also pays 8% sales tax on the purchase, how much does she pay?

   A. $9.6

   B. $120

   C. $126.9

   D. $129.6

2) Will has been working on a report for 3 hours each day, 4 days a week for 3 weeks. How many minutes has Will worked on his report?

   A. 840

   B. 2,160

   C. 36

   D. 1,360

3) James is driving to visit his mother, who lives 147 miles away. How long will the drive be, round–trip, if James drives at an average speed of 70 mph?

   A. 552 minutes

   B. 252 minutes

   C. 522 minutes

   D. 225 minutes

4) In a classroom of 60 students, 39 are female. What percentage of the class is male?

   A. 25%

   B. 35%

   C. 37%

   D. 27%

5) Which of the following is NOT a factor of 36?

   A. 18

   B. 9

   C. 6

   D. 8

AFOQT Subject Test – Mathematics

6) You are asked to chart the temperature during a 6-hour period to give the average. These are your results:

   1 am: 6 degrees   12 am: 26 degrees

   4 am: 11 degrees   10 am: 18 degrees

   5 am: 31 degrees   11 pm: 10 degrees

   What is the average temperature?

   A. 15   C. 17

   B. 21   D. 25

7) During the last week of track training, Emma achieves the following times in seconds: 60, 54, 43, 63, 47, and 70. Her three best times this week (least times) are averaged for her final score on the course. What is her final score?

   A. 48 seconds   C. 66 seconds

   B. 54 seconds   D. 62 seconds

8) How many square feet of tile is needed for 12 feet by 12 feet room?

   A. 144 square feet   C. 124 square feet

   B. 136 square feet   D. 114 square feet

9) With what number must 5.124563 be multiplied in order to obtain the number 51,2456.3?

   A. 10,000   C. 1,000,000

   B. 1,000   D. 100,000

AFOQT Subject Test – Mathematics

10) Emma is working in a hospital supply room and makes $60.00 an hour. The union negotiates a new contract giving each employee a 5% cost of living raise. What is Emma's new hourly rate?

   A. $55 an hour

   B. $53 an hour

   C. $65 an hour

   D. $63 an hour

11) Emily and Lucas have taken the same number of photos on their school trip. Emily has taken 5 times as many photos as Mia. Lucas has taken 36 more photos than Mia. How many photos has Mia taken?

   A. 9

   B. 11

   C. 5

   D. 12

12) Which answer is equivalent to two to the sixth power?

   A. 0.000002

   B. 200,000

   C. 0.64

   D. 64

13) Find the average of the following numbers: 16, 27, 23, 38

   A. 23

   B. 27

   C. 26

   D. 24.3

14) A mobile classroom is a rectangular block that is 23 feet by 19 feet in length and width respectively. If a student walks around the block once, how many yards does the student cover?

   A. 1,244 yards

   B. 437 yards

   C. 84 yards

   D. 124 yards

AFOQT Subject Test – Mathematics

15) What is the distance in miles of a trip that takes 2.8 hours at an average speed of 21.7 miles per hour? (Round your answer to a whole number)

   A. 59 miles

   B. 61 miles

   C. 63 miles

   D. 57 miles

16) The sum of 7 numbers is greater than 100 and less than 150. Which of the following could be the average (arithmetic mean) of the numbers?

   A. 12

   B. 16

   C. 23

   D. 36

17) A barista averages making 16 coffees per hour. At this rate, how many hours will it take until she's made 1,280 coffees?

   A. 85 hours

   B. 60 hours

   C. 80 hours

   D. 120 hours

18) Nicole was making $6.70 per hour and got a raise to $7.14 per hour. What percentage increase was Nicole's raise?

   A. 6%

   B. 5.56%

   C. 6.56%

   D. 5.65%

19) An architect's floor plan uses $\frac{1}{2}$ inch to represent one mile. What is the actual distance represented by $5\frac{1}{2}$ inches?

   A. 11 miles

   B. 12 miles

   C. 5 miles

   D. 6 miles

AFOQT Subject Test – Mathematics

20) A snack machine accepts only quarters. Candy bars cost 95¢, a package of peanuts costs 55¢, and a can of cola costs 25¢. How many quarters are needed to buy one Candy bars, one package of peanuts, and one can of cola?

   A. 12 quarters

   B. 8 quarters

   C. 7 quarters

   D. 4 quarters

21) A writer finishes 240 pages of his manuscript in 60 hours. How many pages is his average per hour?

   A. 4

   B. 6

   C. 3

   D. 9

22) I've got 48 quarts of milk and my family drinks 4 gallons of milk per week. How many weeks will that last us?

   A. 12 weeks

   B. 9.25 weeks

   C. 3 weeks

   D. 1.25 weeks

23) A floppy disk shows 498,686 bytes free and 442,098 bytes used. If you delete a file of size 912,425 bytes and create a new file of size 802,126 bytes, how many free bytes will the floppy disk have?

   A. 1,456,125

   B. 608,985

   C. 110,299

   D. 699,985

24) If $2y + 5y + 11y = -72$, then what is the value of y?

   A. -7

   B. -5

   C. -4

   D. -2

AFOQT Subject Test – Mathematics

25) The hour hand of a watch rotates 30 degrees every hour. How many complete rotations does the hour hand make in 4 days?

A. 8

B. 12

C. 11

D. 15

AFOQT Subject Test – Mathematics

AFOQT Practice Test 2

Mathematics Knowledge

- 25 questions
- Total time for this section: 22 Minutes
- Calculators are not allowed for this test.

Administered Month Year

AFOQT Subject Test – Mathematics

1) Simplify: $2(3x^5)^3$.

   A. $54x^8$

   B. $54x^{15}$

   C. $6x^{15}$

   D. $27x^{15}$

2) What is the perimeter of the triangle in the provided diagram?

   A. 324

   B. 56

   C. 44

   D. 54

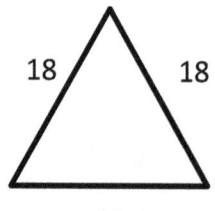

3) If x is a positive integer divisible by 4, and $x < 30$, what is the greatest possible value of x?

   A. 20

   B. 14

   C. 28

   D. 24

4) $(x + 8)(x + 5) = ?$

   A. $x^2 + 40x + 13$

   B. $2x + 8x + 20$

   C. $x^2 + 33x + 40$

   D. $x^2 + 13x + 40$

5) Convert 930,000 to scientific notation.

   A. 9.3×10^5

   B. 9.30×10^{-5}

   C. 9.30×10

   D. 9.30×100

6) Which of the following is an obtuse angle?

   A. 55°

   B. 45°

   C. 150°

   D. 220°

AFOQT Subject Test – Mathematics

7) $8^9 \times 8^4 = ?$

   A. 8^9 C. 8^4

   B. 8^{36} D. 8^{13}

8) What is 6,129.49755 rounded to the nearest tenth?

   A. 6,129.387   C. 6.129

   B. 6,129.4   D. 6,129.380

9) The cube root of 343 is?

   A. 7   C. 777

   B. 77   D. 71,177,577

10) A circle has a diameter of 16 inches. What is its approximate area? (π = 3.14)

    A. 150.66   C. 200.00

    B. 200.96   D. 225.56

11) There are two pizza ovens in a restaurant. Oven 1 burns four times as many pizzas as oven 2. If the restaurant had a total of 135 burnt pizzas on Saturday, how many pizzas did oven 2 burn?

    A. 35   C. 27

    B. 10   D. 17

12) Which of the following is the correct calculation for 8!?

    A. 336 × 5!   C. 8 × 6 × 7!

    B. 120 × 4!   D. 8 × 6 × 6!

AFOQT Subject Test – Mathematics

13) The equation of a line is given as: $y = 4x - 3$. Which of the following points does not lie on the line?

    A. (3, 9)   C. (2, 5)

    B. (−1, −9)   D. (−2, −11)

14) How long is the line segment shown on the number line below?

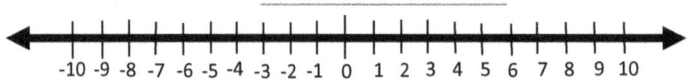

    A. −3   C. 6

    B. −5   D. 9

15) What is the distance between the points (13, 9) and (5, 3)?

    A. 6   C. 12

    B. 16   D. 10

16) $x^2 - 49 = 0$, x could be:

    A. 7   C. 4

    B. 8   D. 12

17) A rectangular plot of land is measured to be 110 feet by 170 feet. Its total area is:

    A. 18,700 square feet   C. 17,800 square feet

    B. 16,700 square feet   D. 17,100 square feet

AFOQT Subject Test – Mathematics

18) Which of the following is NOT a factor of 96?

   A. 24

   B. 6

   C. 12

   D. 13

19) The sum of 3 numbers is greater than 120 and less than 200. Which of the following could be the average (arithmetic mean) of the numbers?

   A. 75

   B. 78

   C. 42

   D. 35

20) One fourth the cube of 8 is:

   A. 512

   B. 64

   C. 128

   D. 32

21) What is the sum of the prime numbers in the following list of numbers?

   24, 11, 5, 14, 36, 23, 39, 49

   A. 29

   B. 27

   C. 32

   D. 39

22) Convert 25% to a fraction.

   A. $\frac{1}{3}$

   B. $\frac{1}{4}$

   C. $\frac{2}{25}$

   D. $\frac{1}{5}$

23) With what number must 9.246895 be multiplied in order to obtain the number 92,468.95?

   A. 1,000

   B. 1,000,000

   C. 10,000

   D. 100,000

AFOQT Subject Test – Mathematics

24) The supplement angle of a 68° angle is:

    A. 125°   C. 92°

    B. 112°   D. 22°

25) 70% of 90 is:

    A. 16   C. 63

    B. 54   D. 49

AFOQT Subject Test – Mathematics

Chapter 13 : Answers and Explanations
Answer Key

In this section, answers and explanations are provided for two AFOQT Math Tests. Review the answers and explanations to learn more about solving questions fast.

AFOQT Math Practice Test

Arithmetic Reasoning -1				Mathematic Knowledge -1				Arithmetic Reasoning -2				Mathematic Knowledge -2			
1	B	16	C	1	C	16	C	1	D	16	B	1	B	16	A
2	B	17	B	2	D	17	C	2	B	17	C	2	D	17	A
3	B	18	B	3	D	18	C	3	B	18	C	3	C	18	D
4	C	19	B	4	C	19	B	4	B	19	A	4	D	19	C
5	A	20	D	5	B	20	D	5	D	20	C	5	A	20	C
6	A	21	D	6	D	21	B	6	C	21	A	6	C	21	D
7	D	22	B	7	A	22	B	7	A	22	C	7	D	22	B
8	D	23	C	8	B	23	C	8	A	23	B	8	B	23	C
9	B	24	A	9	C	24	D	9	D	24	C	9	A	24	B
10	B	25	D	10	B	25	B	10	D	25	A	10	B	25	C
11	C			11	A			11	A			11	C		
12	D			12	C			12	D			12	A		
13	A			13	D			13	C			13	B		
14	B			14	D			14	C			14	D		
15	B			15	B			15	B			15	D		

WWW.MathNotion.Com

AFOQT Subject Test – Mathematics

AFOQT Subject Test – Mathematics

Answers and Explanations

Practice Test 1
Arithmetic Reasoning

1) Answer: B

$\sqrt{144} \times \sqrt{25} = 12 \times 5 = 60$

2) Answer: B

$5\frac{1}{14} - 4\frac{6}{7} = 5\frac{1}{14} - 4\frac{12}{14} = 4\frac{15}{14} - 4\frac{12}{14} = \frac{3}{14} = \frac{3}{14}$

3) Answer: B

$245 - 35 = 210$

$\frac{210}{14} = 15.$

4) Answer: C

To convert a decimal to percent, multiply it by 100 and then add percent sign (%).

$0.034 \times 100 = 3.40\%$

5) Answer: A

$5x - 24x + 19x = -32 \Rightarrow 0 \neq -32 \Rightarrow$ The equation not related to x and you can choose any number.

6) Answer: A

Elise = Mia – 8 $\Rightarrow$ Mia = Elise + 8

Elise = Mason – 6 $\Rightarrow$ Mason = Elise + 6

Mia + Elise + Mason = 110

Now, replace the ages of Mia and Mason by Elise. Then:

Elise + 8 + Elise + Elise + 6 = 110

3Elise + 14 = 110 $\Rightarrow$ 3Elise = 110 – 14

3Elise = 96 $\Rightarrow$ Elise = 32

7) Answer: D

Distance = $speed \times time \Rightarrow$ time = $\frac{distance}{speed} = \frac{175}{25} = 7$

AFOQT Subject Test – Mathematics

(Round trip means that the distance is 175 miles)

The round trip takes 7 hours. Change hours to minutes, then: $7 \times 60 = 420$

8) Answer: D

Change 9 hours to minutes, then: $9 \times 60 = 540$ minutes

$\frac{540}{30} = 18$

9) Answer: B

Since Julie gives 8 pieces of candy to each of her friends, then, then number of pieces of candies must be divisible by 8.

A. $430 \div 8 = 53.75$
B. $584 \div 8 = 73$
C. $326 \div 8 = 40.75$
D. $644 \div 8 = 80.5$

Only choice B gives a whole number.

10) Answer: B

Area of a rectangle = width × length = $52 \times 30 = 1,560$

11) Answer: C

average = $\frac{sum}{total}$

Sum = $10 + 13 + 21 + 27 + 32 + 29 = 132$

Total number of numbers = 6

average = $\frac{132}{6} = 22$

12) Answer: D

The base rate is $35.

The fee for the first 60 visits is: $60 \times 0.40 = 24$

The fee for the visits 61 to 100 is: $40 \times 0.30 = 12$

Total charge: $35 + 24 + 12 = 71$

13) Answer: A

$average = \frac{sum}{total} = \frac{31+42+14}{3} = \frac{87}{3} = 29$

AFOQT Subject Test – Mathematics

14) Answer: B

Find the value of each choice:

A. $2 \times 2 \times 3 \times 3 = 36$

B. $2 \times 2 \times 2 \times 2 \times 2 \times 5 \times 5 = 800$

C. $2 \times 3 \times 17 = 102$

D. $2 \times 3 \times 5 \times 7 = 210$

15) Answer: B

The amount they have = $14.36 + $12.41+ $19.53 = 46.3

16) Answer: C

15 dozen of magazines are 180 magazines: $15 \times 12 = 180$

$180 - 52 = 128$

17) Answer: B

probability $= \frac{desired\ outcomes}{possible\ outcomes} = \frac{4}{4+13+8+7} = \frac{4}{32} = \frac{1}{8}$.

18) Answer: B

1 ton = 2,000 pounds

15 ton = 30,000 pounds

$\frac{75,000}{30,000} = 2.5$

William needs to make at least 2 trips to deliver all the food.

19) Answer: B

$460 - 70 - 150 = 240$

20) Answer: D

Each worker can walk 3 dogs: $15 \div 5 = 3$

7 workers can walk 21 dogs. $7 \times 3 = 21$

21) Answer: D

$54 \div 6 = 9$ hours for one course. $9 \times 35 = 315 \Rightarrow \$315$

22) Answer: B

Write proportion and solve.

AFOQT Subject Test – Mathematics

$\frac{1}{112} = \frac{25}{x} \Rightarrow x = 25 \times 112 = 2,800.$

23) Answer: C

Write a proportion and solve.

$\frac{\frac{1}{7}}{1} = \frac{35}{x}$ x $= \frac{35}{\frac{1}{7}} = 245$

24) Answer: A

6 students did not have to go to summer school.

$60 - 6 = 54$

$\frac{54}{60} = \frac{9}{10}$

25) Answer: D

1 quart = 0.25 gallon

64 quarts = 64 × 0.25 = 16 gallons

then: $\frac{16}{2} = 8$ weeks

AFOQT Practice Test 1
Mathematics Knowledge

1) Answer: C

Only choice C is not equal to 6^3

2) Answer: D

The reciprocal of $\frac{x^4}{81}$ is $\frac{81}{x^4} = \left(\frac{3}{x}\right)^4$

3) Answer: D

If a = 12 then:

$b = \frac{a^2}{6} + 5 \Rightarrow b = \frac{12^2}{6} + 5 = 24 + 5 = 29$

4) Answer: C

$\sqrt[6]{729} = 3$

($3^6 = 2 \times 2 \times 2 \times 2 \times 2 \times 2 = 729$).

5) Answer: B

(r = radius)

Area of a circle = $\pi r^2 = \pi \times (5)^2 = 3.14 \times 25 = 78.50$

6) Answer: D

All angles in a triable add up to 180 degrees.

$90° + 63° = 153°$

$x = 180° - 153° = 27°$

7) Answer: A

$-15a = 165 \Rightarrow a = \frac{165}{-15} = -11$

8) Answer: B

In scientific notation form, numbers are written with one whole number times 10 to the power of a whole number. Number 361,297,580 has 9 digits. Write the number and after the first digit put the decimal point. Then, multiply the number by 10 to the power of 8 (number of remaining digits). Then:

$361,297,580 = 3.6129758 \times 10^8$

AFOQT Subject Test – Mathematics

9) Answer: C

Use Pythagorean Theorem: $a^2 + b^2 = c^2$

$(10)^2 + (24)^2 = c^2 \Rightarrow 100 + 576 = 676 = C^2 \Rightarrow C = \sqrt{676} = 26$

10) Answer: B

Factor of 16: $\{1, 2, 4, 8, 16\}$

Factor of 36: $\{1, 2, 3, 4, 6, 9, 12, 18, 36\}$

Then, factors they have in common is: $\{1, 2, 4\}$

11) Answer: A

$39: \{1, 3, 13, 39\}$.

The rest of choices have the factor of 1 and itself.

12) Answer: C

Use FOIL (first, out, in, last) method.

$(x + 4)(2x + 5) = 2x^2 + 5x + 8x + 20 = 2x^2 + 13x + 20$

13) Answer: D

$11(a - 4) = 33 \Rightarrow 11a - 44 = 33 \Rightarrow 11a = 33 + 44 = 77$

$\Rightarrow 11a = 77 \Rightarrow a = \frac{77}{11} = 7$

14) Answer: D

Use exponent multiplication rule:

$x^a \times x^b = x^{a+b}, (x^a)^b = x^{ab}$

Then: $16^{11} = (4^2)^{11} = 4^{22} = 4^7 \times 4^{3x} = 4^{7+3x}$

$22 = 7 + 3x \Rightarrow 3x = 22 - 7 = 15 \Rightarrow x = 5$

15) Answer: B

Volume = length × width × height

Volume = $11 \times 6 \times 5 \Rightarrow$ Volume = 330 cm^3

16) Answer: C

Slope of a line: $\frac{y_2 - y_1}{x_2 - x_1} = \frac{rise}{run}$

$\frac{y_2 - y_1}{x_2 - x_1} = \frac{8 - 2}{5 - 4} = \frac{6}{1} = 6$

WWW.MathNotion.Com

AFOQT Subject Test – Mathematics

17) Answer: C

The straight line is 180 degrees. Then:

$\angle a° = 180° - 58° = 122°$.

18) Answer: C

To factor the expression $x^2 - 4x - 32$, we need to find two numbers whose sum is -4 and their product is -32.

Those numbers are 8 and -4. Then: $x^2 - 4x - 32 = (x-8)(x+4)$

19) Answer: B

The area of the non-shaded region is equal to the area of the bigger rectangle subtracted by the area of smaller rectangle.

Area of the bigger rectangle = $12 \times 8 = 96$

Area of the smaller rectangle = $5 \times 4 = 20$

Area of the non-shaded region = $96 - 20 = 76$

20) Answer: D

Diameter $D = 2r \Rightarrow 10 = 2r \Rightarrow r = 5$

Area $= \pi r^2 \Rightarrow A = \pi(5)^2 \Rightarrow A = 25\pi$

21) Answer: B

Diameter $D = 2r \Rightarrow 16 = 2r \Rightarrow r = 8$

Circumference $= 2\pi r \Rightarrow C = 2\pi r \Rightarrow C = 16 \times 3.14 = 50.24$

22) Answer: B

$\begin{cases} 5x - 3y = -4 \\ (-3x + y = 4) \times 3 \end{cases} \Rightarrow \begin{cases} 5x - 3y = -4 \\ -9x + 3y = 12 \end{cases} \Rightarrow -4x = 8 \Rightarrow x = -2$

$-3x + y = 4 \Rightarrow -3(-2) + y = 4 \Rightarrow y + 6 = 4 \Rightarrow y = 4 - 6 \Rightarrow y = -2.$

23) Answer: C

$\sqrt{18} = \sqrt{9} \times \sqrt{2} = 3\sqrt{2}$

$\sqrt{72} = \sqrt{36} \times \sqrt{2} = 6\sqrt{2}$

$3\sqrt{2} \times 6\sqrt{2} = 3 \times 6 \times 2 = 36.$

AFOQT Subject Test – Mathematics

24) Answer: D

An acute angle is an angle of greater than 0 degrees and less than 90 degrees. Only choice a is an obtuse angle.

25) Answer: B

Factorial:

$$n! = 1 \times 2 \times 3 \times \ldots \times n$$

$$n! = n(n-1)(n-2)(n-3)!$$

$$\frac{8!}{5!} = \frac{(8 \times 7 \times 6) \times 5!}{5!} = 336.$$

AFOQT Subject Test – Mathematics

Answers and Explanations

Practice Test 2
Arithmetic Reasoning

1) Answer: D

$25\% \times 160 = \frac{25}{100} \times 160 = 40$; The coupon has $40 value.

Then, the selling price of the sweater is $120 (160 − 40 = 120).

Add 8% tax, then: $\frac{8}{100} \times 120 = 9.6$ for tax

then: $120 + 9.6 = 129.6$

2) Answer: B

4 weeks = 12 days. Then: $12 \times 3 = 36$ hours

$36 \times 60 = 2{,}160$ minutes

3) Answer: B

Distance = speed × time ⇒ time = $\frac{distance}{speed} = \frac{147+147}{70} = 4.2$

(Round trip means that the distance is 294 miles)

The round trip takes 4.2 hours. Change hours to minutes, then:

$4.2 \times 60 = 252$

4) Answer: B

$60 − 39 = 21$ male students

$\frac{21}{60} = 0.35$

Change 0.35 to percent ⇒ $0.35 \times 100 = 35\%$

5) Answer: D

The factors of 36 are: {1, 2, 3, 4, 6, 9, 12, 18, 36}

8, is not a factor of 36.

6) Answer: C

average = $\frac{sum\ of\ values}{number\ of\ values}$

Sum = $6 + 11 + 31 + 26 + 18 + 10 = 102$

AFOQT Subject Test – Mathematics

Total number of numbers = 6

$\frac{102}{6} = 17$

7) Answer: A

Emma's three best times are 43, 47, and 54.

The average of these numbers is: $average = \frac{sum\ of\ values}{number\ of\ values}$

Sum = 43 + 47 + 54 = 144

Total number of numbers = 3

$average = \frac{144}{3} = 48.$

8) Answer: A

The area of 12 feet by 12 feet room is 144 square feet.

12 × 12 = 144

9) Answer: D

5.124563 × 100,000 = 51,2456.3

10) Answer: D

5 percent of 60 is: $60 \times \frac{5}{100} = 3$

Emma's new rate is 63. (60 + 3 = 63).

11) Answer: A

Emily = Lucas

Emily = 5 Mia ⇒ Lucas = 5 Mia

Lucas = Mia + 36

then: Lucas = Mia + 36 ⇒ 5 Mia = Mia + 36

Remove 1 Mia from both sides of the equation. Then: 4Mia = 36 ⇒ Mia = 9

12) Answer: D

$2^6 = 2 \times 2 \times 2 \times 2 \times 2 \times 2 = 64$

13) Answer: C

Sum = 16 + 27 + 23 + 38 = 104

$average = \frac{104}{4} = 26$

AFOQT Subject Test – Mathematics

14) Answer: C

Perimeter of a rectangle = 2 × length + 2 × width =

2 × 23 + 2 × 19 = 46 + 38 = 84

15) Answer: B

$Speed = \frac{distance}{time}$

$21.7 = \frac{distance}{2.8} \Rightarrow distance = 21.7 \times 2.8 = 60.76$

Rounded to a whole number, the answer is 61.

16) Answer: B

Let's review the choices provided and find their sum.

a. 12 × 7 = 84

b. 16 × 7 = 112 ⇒ is greater than 100 and less than 150

c. 23 × 7 = 161

d. 36 × 7 = 252

Only choice B gives a number that is greater than 100 and less than 150.

17) Answer: C

$\frac{1 \ hour}{16 \ coffees} = \frac{x}{1,280} \Rightarrow 16 \times x = 1 \times 1,280 \Rightarrow 16x = 1,280 \Rightarrow x = 80$

It takes 80 hours until she's made 1,280 coffees.

18) Answer: C

$percent \ of \ change = \frac{change}{original \ number}$,

7.14 – 6.70 = 0.44

$percent \ of \ change = \frac{0.44}{6.70} = 0.0656 \Rightarrow 0.0656 \times 100 = 6.56\%$

19) Answer: A

Write a proportion and solve.

$\frac{\frac{1}{2} inches}{5.5} = \frac{1 \ mile}{x}$.

Use cross multiplication, then: $\frac{1}{2}x = 5.5 \rightarrow x = 11$.

AFOQT Subject Test – Mathematics

20) Answer: C

One candy bars costs 95¢ and a package of peanuts cost 55¢ and a can of cola costs 25¢. The total cost is:

$95 + 55 + 25 = 175$, 175 is equal to 7 quarters.

$7 \times 25 = 175$

21) Answer: A

$240 \div 60 = 4$

22) Answer: C

1 quart = 0.25 gallon

48 quarts = $48 \times 0.25 = 12$ gallons

then: $\frac{12}{4} = 3$ weeks

23) Answer: B

The difference of the file added, and the file deleted is:

$912,425 - 802,126 = 110,299$

$498,686 + 110,299 = 608,985$

24) Answer: C

$2y + 5y + 11y = -72 \Rightarrow 18y = -72 \Rightarrow y = -\frac{72}{18} \Rightarrow y = -4$

25) Answer: A

Every day the hour hand of a watch makes 2 complete rotation. Thus, it makes 8 complete rotations in 4 days.

$24 \times 30 = 720 \Rightarrow 720 \div 360 = 2$

$2 \times 4 = 8$

AFOQT Subject Test – Mathematics

AFOQT Practice Test 2

Mathematics Knowledge

1) Answer: B

$2(3x^5)^3 \Rightarrow 2 \times 3^3 \times x^{15} = 54x^{15}$

2) Answer: D

Perimeter of a triangle = side 1 + side 2 + side 3 = 18 + 18 + 18 = 54

3) Answer: C

From the choices provided, 20, 28 and 24 are divisible by 4. From these numbers, 28 is the biggest.

4) Answer: D

Use FOIL (First, Out, In, Last) method.

$(x + 8)(x + 5) = x^2 + 5x + 8x + 40 = x^2 + 13x + 40$

5) Answer: A

In scientific notation form, numbers are written with one whole number times 10 to the power of a whole number. Number 930,000 has 6 digits. Write the number and after the first digit put the decimal point. Then, multiply the number by 10 to the power of 5 (number of remaining digits). Then:

$930,000 = 9.3 \times 10^5$

6) Answer: C

An obtuse angle is an angle of greater than 90° and less than 180°.

7) Answer: D

Use exponent multiplication rule: $x^a \cdot x^b = x^{a+b}$

Then: $8^9 \times 8^4 = 8^{13}$

8) Answer: B

6,129.49755 rounded to the nearest tenth equals 6,129.5

(Because 6,129.49 is closer to 6,129.5 than 6,129.4)

9) Answer: A

$\sqrt[3]{343} = 7$

AFOQT Subject Test – Mathematics

10) Answer: B

Diameter = 16

then: Radius = 8

Area of a circle = πr^2 ⇒ A = $3.14(8)^2$ = 200.96

11) Answer: C

Oven 1 = 4 oven 2

If Oven 2 burns 4 then oven 1 burns 2 pizzas.

Oven 1 + oven 2 = 5 burn pizzas.

135 ÷ 5 = 27 oven 2

135 − 27 = 108 (4 × 27) oven 1

12) Answer: A

$n! = n(n-1)(n-2)(n-3)(n-4)!$

$8! = 8 \times 7 \times 6 \times 5! =$

13) Answer: B

Let's review the choices provided. Put the values of x and y in the equation.

A. (3, 9)   ⇒ $x = 3$ ⇒ $y = 9$ This is true!

B. (−1, −9) ⇒ $x = -1$ ⇒ $y = -7$ This is not true!

C. (2, 5)   ⇒ $x = 2$ ⇒ $y = 5$ This is true!

D. (−2, −11) ⇒ $x = -2$ ⇒ $y = -11$ This is true!

14) Answer: D

6 − (−3) = 9

15) Answer: D

Use distance formula: $d = \sqrt{(x_1 - x_2)^2 + (y_1 - y_2)^2} = \sqrt{(13 - 5)^2 + (9 - 3)^2}$

$\sqrt{64 + 36} = \sqrt{100} = 10$

16) Answer: A

$x^2 - 49 = 0$ ⇒ $x^2 = 49$ ⇒ x could be 7 or −7.

17) Answer: A

Area of a rectangle = width × length = 110 × 170 = 18,700

AFOQT Subject Test – Mathematics

18) Answer: D

factor of 96 = {1, 2, 3, 4, 6, 8, 12, 16, 24, 32, 48, 96}

13 is not a factor of 96.

19) Answer: C

Let's review the choices provided.

A. $75 \times 3 = 225$

B. $78 \times 3 = 234$

C. $42 \times 3 = 126$

D. $35 \times 3 = 105$

From choices provided, only 126 is greater than 120 and less than 200.

20) Answer: C

The cube of $8 = 8 \times 8 \times 8 = 512$

$\frac{1}{4} \times 512 = 128$

21) Answer: D

From the list of numbers, 11, 5, and 23 are prime numbers. Their sum is:

$11 + 5 + 23 = 39$

22) Answer: B

$25\% = \frac{25}{100} = \frac{1}{4}$

23) Answer: C

Number 9.246895 should be multiplied by 10,000 in order to obtain the number 92,468.95

$9.246895 \times 10,000 = 92,468.95$

24) Answer: B

Two Angles are supplementary when they add up to 180 degrees.

$112° + 68° = 180°$

25) Answer: C

$\frac{70}{100} \times 90 = 63$

"End"

www.ingramcontent.com/pod-product-compliance
Lightning Source LLC
Chambersburg PA
CBHW080438110426
42743CB00016B/3205